FLY

FLY

The Unsung Hero of Twentieth-Century Science

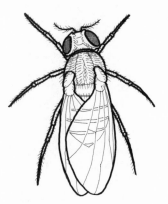

Martin Brookes

An Imprint of HarperCollins*Publishers*

Illustrations by Mari Estrella

FIRST EDITION

Library of Congress Cataloging-in-Publication Data
Brookes, Martin, 1967–
Fly : the unsung hero of twentieth-century science / by
Martin Brookes.—1st ed.
p. cm.
"Published in Great Britain as: Fly : an experimental life by
Weidenfeld & Nicolson."—Copr. p.
Includes bibliographical references.
ISBN 0-06-621251-0 (hardcover)
1. Genetics—History. 2. Drosophila melanogaster. I. Title.
QH428 .B766 2001
576.5'09—dc21
2001023845

01 02 03 04 05 ❖/RRD 10 9 8 7 6 5 4 3 2 1

for Jennifer

ACKNOWLEDGMENTS

I want to thank the following people for the help and support they provided during the writing of this book: Jennifer Brady, Jenny Bangham, Tracey Chapman, David Concar, Alice Hunt, Owen Rose, and Peter Tallack.

CONTENTS

THE FLY WHO CAME IN FROM
THE COLD

Inside the cage, John and Yoko were going through the mating ritual. John was the more active of the two, vibrating various bits of his anatomy at physically implausible speeds, while Yoko looked on impassively. Peering through the Plexiglas walls, we whispered laddish encouragement, urging John to make his move. When eventually he did, climbing on top of his partner from the rear, the studious silence of the genetics class was interrupted by our loud chorus of orgasmic cheers.

I spent most of that afternoon with two university friends creating daft nicknames for our captive fruit flies and paying little attention to the science. "John and Yoko," "Sid and Nancy," and "Charles and Di" seemed preferable to the dry and prosaic *Drosophila melanogaster*. Dozens of star-studded couples passed under our impatient adolescent gaze. Sometimes the flies would sit motionless at opposite ends of their enclosure. Bored and frustrated, we would flick at their cages, willing them into doing something worth watching.

It was difficult to take the fruit fly seriously. Like all insects, it had a head, a thorax, an abdomen, and six delicate legs. It also had wings; two of them. But with all this presented in a body less than half the size of a grape seed, here was an animal crying out to be ignored. You could squash a hundred of them without noticing, and I did. On my own, purely subjective, scale of animal aesthetics, the fruit fly ranked pretty low; respectably higher than the flatworm, but some way below the dog whelk.

Even among its evolutionary relatives, the fruit fly hardly seemed to stand out. It lacked the ghoulish charm of distant cousins like screwworm flies, which laid their eggs in the genitals, mouth, and nose of their hapless mammalian victims. It had none of the infectious stealth of disease-mongers like mosquitoes, with their incumbent coterie of parasitic hangers-on. It didn't even have any annoying agricultural habits, unlike the notorious med-fly (also known, confusingly, as a "fruit fly"), which grabbed head-lines by destroying citrus crops in California and Europe.

As far as I was concerned, screwworm flies, mosquitoes, medflies, and the like were the real party animals: flies that evolution had blessed with intrinsically interesting lives. The fruit fly, on the other hand, seemed like an early-to-bed-with-a-cup-of-hot-cocoa sort of fly.

But my feelings about the fly soon changed. After finishing my degree, I went looking for a Ph.D. in evolutionary biology, and was confronted with a baffling choice of projects and organisms. At the time, I was more concerned with the organism than the details of the science. My main priority was to work on a "proper" animal, something brightly colored with fur or feathers living in a remote part of the Amazon. Unfortunately, this seemed to be a priority shared by most of my peers. So, in the end, I had to settle for what I could get—a project on a small species of moth in South Wales.

The project lacked the glamour I had yearned for, but later I realized that this could be a blessing in disguise. Choosing a glamorous Ph.D. could mean a free ticket to the tropics. But more often than not, you would return, three years later, with an empty notebook, a bad dose of malaria, and a scientific career in tatters. Once you had learned the language, established a base camp, and set up your tripod, it was time to come home. You could always pick out these choice victims at academic conferences. They were the ones wearing glazed expressions over their suntans.

But there were others who stood out from the conference crowd: young, self-confident individuals whose demeanor suggested they were going places. These were people for whom public speaking seemed to hold no fears. They gave talks that translated each short scientific life into one long success story. They collected new facts like a bee collected pollen. And they had their work routinely published in the distinguished pages of *Nature* and *Science*. They came from all corners of the globe but were united by a common bond. Who were these people?

They were the ones who had chosen to work with fruit flies.

If my brief stint in academia taught me anything, it was that my system of animal aesthetics was completely incompatible with the practical, temporal, and financial constraints of biological research. The animals I'd dismissed as irrelevant—small insects, for example—excelled as research tools. And the animal I'd dismissed as most irrelevant—the fruit fly—stood head and thorax above all others. The fruit fly came with all the attributes of other small insects. But it also came with something else: a long and distinguished scientific history.

The fly made its official laboratory debut in 1900, under the watchful eye of Harvard University professor William Castle.

In all truth, the crossing of the laboratory threshold was very much a nonevent. Castle needed a study organism for one of his embryology students. The fruit fly seemed like a cheap and cheerful option, so a few ripe grapes were left on a windowsill, and any flies that took the bait were brought inside.

The fly was just one of many new experimental animals to be tried and tested during the tail end of the Victorian era, as biology underwent a major revamp. For most of the nineteenth century, biology had been dominated by the naturalist philosophy. Naturalists believed that uncovering biological truths depended on meticulous observations of life in its natural context. As a consequence, biology was characterized by a near obsession with descriptive detail. From a tiny hair on a beetle's bottom to a family of fleas in a kangaroo's crotch, nothing was too trivial to document.

But as the nineteenth century wore on, the naturalists came under increasing attack from a new generation of biologists who viewed life with more materialistic and mechanistic eyes. The study of life, they argued, was best approached not only by describing what exists, but through carefully controlled experiments and manipulations. Eventually, tradition gave way to this new wave of experimental biology. By the turn of the century, natural history was in serious decline.

Liberated from the straitjacket of natural history, biology began to diversify into specialized disciplines such as animal behavior, evolution, and physiology. Eager to test a glut of new ideas, biologists went looking for organisms well suited to an experimental life indoors. The fruit fly proved a champion.

But the fly was not an obvious candidate for laboratory superstar. Its small size immediately put it at a disadvantage in a Victorian society that valued bigger and bolder beasts as potent symbols of biological kudos. Those animals held in the highest regard by the Victorian middle classes—dogs, cats, pigeons, even rats and mice—were, ideally, the ones to be seen with.

Given the climate of animal snobbery, it may seem astonishing that something as "low class" as the fruit fly ever made it across the laboratory threshold. But Victorian values meant nothing to a hardened itinerant like the fly. A shameless self-promoter, it seemed to demand human attention. It loitered around dustbins parked conveniently close to kitchen doors. It established impromptu "love-ins" in half-empty lunch boxes, carelessly left open on summer lawns. And it made dangerous explorations deep into the heart of Victorian drawing rooms in search of warmth and abandoned fruits. A catholic taste in food allowed the fly to exploit the free banquets unwittingly laid on by *Homo sapiens.* Wherever fruit and vegetables were stored, preserved, fermented, or simply left to rot, fruit flies followed.

The fly's early years in the laboratory were productive, if unremarkable. There was certainly nothing to suggest that it was destined for great things. For much of the time, the fly found itself in the clumsy hands of gawky adolescents. With the move toward experimental biology, practical work and research projects were becoming an increasing part of a young person's biological education, and there was a real need for an animal that could fulfill the role of laboratory stooge. With its "live fast, die young" philosophy, the fly was supremely adaptable to the tight schedules of the academic calendar, and tailor-made for research against the clock.

Their small size and easygoing habits meant that flies were cheap to house and cheap to feed. A half-pint milk bottle with a piece of rotting banana would keep two hundred flies happy for a fortnight. They were also easy to breed, with each female laying several hundred eggs. What is more, the flies got on with life: birth, sex, and death were all wrapped up inside a few tumultuous weeks. In short, fruit flies did pretty much what other animals did, only cheaper and faster.

News of the fly traveled at a modest pace among William Castle's academic network. By 1907, the fly had founded additional laboratory colonies at Indiana University in Bloomington, Bryn Mawr College in Pennsylvania, and Cold Spring Harbor Laboratory in New York State. But it was at Columbia University in New York City where the fly's laboratory career really took off. It was there, in 1909, that the fly first revealed its talent for the unexpected. A spontaneous change in eye color caught the attention of zoology professor Thomas Hunt Morgan. It was a small change with a giant impact.

Before the fly, ideas about biological inheritance were a strange amalgam of crackpot hypotheses, myth, and superstition. But at Columbia, the subject was rapidly transformed into a coherent science as Morgan and the fly began to lay the foundations of modern genetics. Morgan proved that the physical basis of heredity in fruit flies lay within threadlike structures, called chromosomes, inside the fly's cells. In addition, he showed that each of these chromosomes consisted of a long list of hereditary instructions—genes—that could be rearranged by reproduction into new and unique combinations.

What was true in fruit flies turned out to be true in other animals, including ourselves: genes and chromosomes are a common hereditary currency. Very rapidly, the fly became *the* experimental animal of choice for any self-respecting geneticist. In 1910–11, there were only five laboratories in the United States and two in Europe carrying out research on the fly. By

1936–37, the fly was taking up space in twenty-six U.S. laboratories and twenty laboratories in Europe.

For thirty years, the fly was at the cutting edge of genetics research. When the first genetic maps, showing the linear order of genes along chromosomes, were being constructed, it was fly genes that were being drawn. When chromosomes were being bombarded with X rays in order to understand the physical nature of genetic mutations, it was fly chromosomes that were on the receiving end.

The importance of this early work cannot be overestimated. Techniques for pinpointing genes for human disease depend on genetic mapmaking principles first established with the fly. It was fruit fly research that opened our eyes to the dangers of radiation to human health. In fact, everything in modern genetics, from gene therapy to cloning to the Human Genome Project, is built on the foundations of early-twentieth-century fruit fly research.

The new genetics soon filtered into other areas of biology. In the 1930s, for example, Russian-born biologist Theodosius Dobzhansky pioneered the merger of genetics and Darwinian evolution to form a new science—the imaginatively titled evolutionary genetics. Genetics gave evolutionary biology a scientific credibility it had previously been lacking, and fruit flies were once again at the forefront. Dobzhansky dismissed the notion that evolution is always a long, drawn-out affair, resistant to scientific inquiry, when he showed that wild fruit fly populations can evolve over a matter of months.

Alas, the fly's scientific momentum could not be sustained, and the middle years of the twentieth century brought a shift in its scientific fortunes. While the fly did not exactly disappear from the research scene, it found itself usurped by a new generation of laboratory pioneers.

In some ways the fly became a victim of its own success. In helping to identify genes as the fundamental units of heredity, it had taken this particular brand of biology as far as it could go. The next logical step was to ask what genes were made of and how genes worked. These were questions of biochemistry and molecular biology, questions that demanded an altogether different form of laboratory life. What was needed was biology at its most basic. It was time for the fly to step aside and make way for viruses, bacteria, yeasts, and molds.

Over the next forty years, this quartet of biological simpletons became the new stars of the scientific show, participating in a clutch of important and influential discoveries: the revelation that DNA is the genetic material; the discovery of the double-helix structure of the DNA molecule; the realization that DNA was a code that does not exert its effects directly, but through chemicals called proteins; the deciphering of the DNA code; and, perhaps most fascinating of all, the discovery that the genetic code is universal. Bacteria, fruit flies, cabbages, and humans may have their differences, but it became clear that these, and the millions of other species on our crowded planet, are all cut from the same chemical cloth.

Among this new generation of laboratory upstarts, it was the bacteria that truly won over the hearts and minds of biologists. Their remarkable ability to exchange bits of DNA with one another in a variety of eccentric ways was a biologist's dream. It hinted at ways of artificially manipulating genes, of moving them around from one organism to another. In short, it laid the foundations for genetic engineering.

Ironically, these developments worked in the fruit fly's favor.

The fly's scientific sex appeal may have been diminished by a new breed of laboratory organisms, but the bacteria-led revolution in molecular biology eventually paved the way for the fruit fly's renaissance.

In the 1970s, the fly was reborn and remodeled as the darling of developmental biology. The question of how a fertilized egg can develop into a fully grown organism had puzzled biologists for centuries. Suddenly, the fly was coming up with answers. And once again, the rules of the game turned out to be more than a fruit fly concern. The blueprint for a fruit fly's body is a useful guide to body building in general. Even the study of human embryonic development has learned a thing or two from flies.

Since the 1970s, a broad church of biologists has been drawn toward the fly's downbeat and homely charms. Success has bred success, and today there are few areas of biology that have not felt the fruit fly's influence. You will find it being used in the search for cancer cures; as an early warning system for global warming and climate change; in the study of neurodegenerative disorders such as Alzheimer's disease and Huntington's chorea; and in understanding the genetics of alcohol and drug addiction, sleeping disorders, and jet lag.

In fact, the fly is uncovering answers to some of the most fundamental questions in biology. How do genes link one generation to the next? How does an egg—a single cell—become an adult, with its billions of different cells? How do we learn and memorize information? Why are males and females in perennial conflict over sex? Why do we age, and can we prevent it? And how do new species evolve?

The laboratory stalwart *Drosophila melanogaster* remains the star of the show. But the fruit fly story goes way beyond the confines of the laboratory. Worldwide, a supporting cast of some two thousand *Drosophila* species has made a significant contribution to the fruit fly's scientific legacy.

Of course, not everything the fruit fly has touched has turned to gold. Far from it. Most biological research is actually rather dull. Take my own research, for example. In pursuit of my Ph.D., I spent four years working out how far a species of moth moves from birth until death. (The answer, in case anybody is interested, is 3.9 meters. Approximately.) Only a tiny fraction of new research actually moves science forward. The rest is merely repetition of what has gone before, with perhaps one or two tweaks here and there to try to make it look less obvious.

Thus far, about a hundred thousand scientific papers have been published on the fly. It is an enormous quantity, and a testament to the fly's enduring and widespread popularity. But of this number I would estimate that only 5 percent have been read by more than a dozen people. The rest remain largely unread, and are useful only in the sense that they provide nourishment for hungry book lice. Still, the 5 percent fruit of fruit fly research that is good is very good indeed. It is the 5 percent that has transformed twentieth-century biology.

All the more remarkable, then, that outside academic circles the fly's public image remains as low as ever. Even when the fruit fly's name does get a mention, the adjective "humble" or "lowly" is usually stuck in front of it. Publicly at least, we seem reluctant to accept that this tiny creature can teach us anything, let alone anything about ourselves. In this book, I intend to set the record straight.

Let's put things in perspective. Apart from the odd weirdo, most biologists do not study the fly solely because they are in love with the minutiae of fruit fly biology. They study the fly in the hope that it will offer pointers to a broader biological picture, one that embraces a wide range of organisms, including ourselves. That the fly is still going strong after all these years is proof positive that, in many instances, these hopes have been realized.

So resonant is the fruit fly's experimental life that this book can be read as a universal story of birth, school, work, death, and a few points in between. In each chapter, I've used fruit fly biology to illustrate a successive stage in life, mapping out the major biological landmarks on the eternal loop of birth and death. From genetics to embryonic development, from learning to sex, from death of the individual to the birth of a new species, the fruit fly has become a window on our biological world.

This dip into the past hundred years of the fruit fly's experimental life is not intended to be an exhaustive survey. With a hundred thousand papers already in print, and more being added each day, you'd have to be insane—or an academic—to embark on such a venture. Rather, the book offers a glimpse of how one short life has helped define the boundaries of our biological knowledge.

Here's to John and Yoko, biological icons of the twentieth century.

1

THE LEGACY OF A LIFETIME

Two red, compound eyes stared out from the undergrowth of the professor's wiry beard. The fly clung to the shaft of a stout, graying chin hair, while digestive juices worked chemical magic on a belly full of banana goo. After a statuesque half-hour, the fly began a prolonged bout of grooming—preening, stroking, and caressing difficult-to-reach parts of its anatomy that didn't really need to be preened, stroked, or caressed.

Now feeling in more of a flying mood, the fly launched itself into Professor Morgan's facial atmosphere. But instead of heading up and away, it followed a shallow clockwise orbit of Morgan's face. Two mutually opposing forces held the fly on its trajectory. On the one hand, the nerve endings in its distended stomach were telling its primitive brain to get the hell out. And on the other, the sensory cells on its antennae were demanding a return to the morsel of rotting banana ensnared in the hooks and barbs of the professor's facial hair. This neuronal indecision was to prove costly.

As the fly passed below Morgan's nose for a second time, it felt a huge tug from behind and promptly disappeared up the professor's right nostril. Disorientated within a dark jungle of spiky nasal hair and mucous swamp, the fly struggled to find an exit. Morgan, meanwhile, had already reached for his hand-kerchief. As he gave his nose a hearty blow, the fly approached supersonic speed. Death came quickly, smeared across the fibers of a smooth cotton tartan.

Morgan sat down and surveyed the clutter of books and bottles on his desk. From the adjoining room, the sound of self-confident science drifted in through the open door. He considered joining the argument, but a bottle with a loose stopper caught his attention and he reached over to pick it up.

With the stopper secured, he held the bottle up to the light to get a better view of the Lilliputian world inside. The flies were going about their daily business. Some were attempting to mount their neighbors. Others were already coitally entwined. A few individuals stood alone on the periphery, apparently disillusioned with the whole mating game. Morgan marveled at how the flies seemed so oblivious to the world outside, so totally absorbed in their basic rituals. He put the bottle down, cleared a space for himself, and began to sketch the outline of his next important manuscript.

Thomas Hunt Morgan made the fruit fly famous. Between 1910 and 1915, he and his research team at New York's Colum-bia University bred billions of flies. To those on the outside, these breeding bonanzas must have seemed like experiments in orgiastic madness. But there was method in the madness. This period was as productive for Morgan as it was for the flies. During these six years, he and his team painstakingly articulated the basics of modern genetics.

The story of how Morgan met up with the fruit fly is a tale of two opportunists, one tall and bearded, with an obsession for experimental science, the other small and bristled, with an obsession for experimental sex. United by their mutual and unceasing passion for productivity, it was a marriage made in heaven, and consummated in the laboratory.

American history also played its part in this meeting of man and beast. The slave trade provided the transport that delivered the fruit fly to the shores of the New World. It was slavery, too, that sent America headlong into civil war. And it was the war's political conclusion that created an academic culture in which a young, intelligent, and inquiring mind could thrive. Morgan's meeting with the fly was no fluke. It was destiny, in the great American tradition. History set them rolling on a collision course that would culminate in their revolutionary encounter in turn-of-the-century New York.

The American Civil War was a watershed for American biology. Before the war, biology was simply an extension of theology. The purpose of biological study was to observe the intricacies of God's grand design. Natural history museums functioned as surrogate cathedrals, disseminating the Lord's message to the public via the natural world's vast repertoire of divine shape and form.

But in the immediate aftermath of the war, amid a climate of political and cultural reform, American academics looked to Europe, and Germany in particular, for a new biological philosophy. In the light of Charles Darwin's new evolutionary theory, European biologists were viewing the natural world in a different light. Biology was shrugging off its theological cloak as secular and utilitarian ideas took hold.

With natural history no longer a pious search for the patterns in God's handiwork, biology became a whole new world of inquiry. The United States followed the European example by moving biology out of the museum and into newly built academic departments and research institutions. This new emphasis on research was eagerly adopted by progressive universities such as Johns Hopkins, Harvard, Chicago, Michigan, and Cornell.

These changes brought with them a renewed interest in experimental biology. Experimental biologists had been lurking in the shadows of mainstream biology since the late seventeenth century, stalking dark, damp basements, and doing unpleasant things to frogs. But the influence of the experimentalists within biology had always been suppressed by the naturalists, who viewed them with suspicion and disdain. Experiments, naturalists argued, could only ever create narrow and oversimplified interpretations of the natural world.

But by the latter part of the nineteenth century, the naturalist viewpoint was coming under pressure from the new wave of biological materialism. The discovery of novel biological techniques was making experimental research in biology a viable and practical alternative. For the first time, high-powered microscopes and chemical stains and dyes were lighting up the inner architecture of cells. Specialized cutting tools were producing accurate sections of animal and plant tissues. Electrical devices were providing precise measurements of physiological changes. And the emergence of anesthesia was making the use of animals both more accessible and humanely acceptable.

It was in this vibrant and progressive climate that Morgan began his postgraduate studies in zoology. In 1886, aged twenty, he enrolled at Johns Hopkins University in Baltimore, one of the new research-orientated institutions that had sprung up after the Civil War. The science of life seemed full of new prospects and possibilities. Arguments between opposing biological traditions were still plentiful and rivalries strong, but there was, at least, the overwhelming sense that biology had been roused from a stultifying complacency.

Morgan's interest in biology had been obvious from an early age, but there was little in his background to suggest that a career in biological research was in the cards. An illustrious family history, however, suggested that he was destined for great things, whatever his chosen path. His father, Charlton Hunt Morgan, had been an American consul in Sicily and had assisted Garibaldi's struggle for Italian independence. John Hunt Morgan, an uncle, was a famous general in the Confederate army and the leader of a guerrilla band known as "Morgan's Raiders." John Wesley Hunt, a great-grandfather, and a prototype Richard Branson, amassed a fortune growing hemp, breeding racehorses, and founding a railroad. And perhaps most famous of all, Francis Scott Key, another of Morgan's great-grandfathers, was the American lawyer and poet who wrote "The Star-Spangled Banner."

It is ironic, considering the direction his future career would take, that Morgan's first forays into biological research were very much in the naturalist tradition. The subject of his Ph.D. thesis—the classification of sea spiders—was supremely dull, even by the dour standards of nineteenth-century natural history.

Sea spiders themselves are engaging little creatures. They live on the seafloor, often at great depths. Unusually, their gonads are contained in their legs and connect with the surface via a myriad of tiny pores. Come the mating season, the legs work like garden sprinklers, spraying fountains of gametes into the ocean.

Unfortunately for Morgan, all the best bits of sea spider biology lay beyond the scope of his own studies. His task was to classify the sea spiders, to pinpoint their location in the tree of life. Sea spiders have characteristics of both spiders and crustaceans (lobsters and crabs) and their taxonomic status has always been a contentious issue for the two or three people in the world who are interested in this sort of thing. Morgan focused his attention on the distinguishing features of the sea spider embryo. After thousands of lonely hours spent staring down a microscope, he concluded that sea spiders were, in fact, more spider than crab.

Despite the dry and descriptive nature of his own work, his time at Johns Hopkins brought Morgan into contact with a wide range of biological perspectives. Within the zoology department, there were plenty of opportunities to rub shoulders with biologists of an experimental persuasion, and it was through these interactions that Morgan's appetite for experimental science was born.

In 1891, and with his Ph.D. completed, Morgan left Johns Hopkins to take up his first professional post, as associate professor of biology at Bryn Mawr College, one of a growing number of women's colleges to appear in the United States after the end of the Civil War.

At Bryn Mawr, Morgan joined up with Jacques Loeb, a German-born physiologist with a strong background in experimental biology. Loeb made a deep impression on his younger colleague, and encouraged Morgan to visit European universities and laboratories, to gain valuable experience in new experimental techniques and research methods.

Morgan learned a great deal from these European jaunts. He particularly enjoyed his trips to the Stazione Zoologica, a marine-biology laboratory in Naples and a Mecca for visiting biologists from all over the world. In 1896, his enthusiasm was palpable when he wrote:

> At the Naples Station are found men of all nationalities. Investigators, professors, privatdocents, assistants and students come from Russia, Germany, Austria, Italy, Holland, England, Belgium, Switzerland and "America"—men of all shades of thought and all sorts of training. The scene shifts from month to month like the turning of a kaleidoscope. No one can fail to be impressed and to learn much in the clash of thought and criticism that must be present where such diverse elements come together.

Frequent trips to the Naples station exposed Morgan to experimental biology's huge potential and influenced the direction of his own research career. As the 1890s progressed, his interests diversified. Soon he was dabbling with almost any aspect of biology that took his fancy, so long as it was amenable to experimental inquiry.

One problem to which he devoted more time than most was regeneration—the ability of animals to grow back amputated bits of their bodies. Morgan came to the problem of regeneration through an interest in embryonic development, considering the two phenomena to be flip sides of the same coin. Both involve the recruitment of cells to specific tasks at specific places. The biological signals that direct a leg stump to develop into the bone, muscle, and skin of a regenerated limb must, he believed, be similar to those that direct cells within a growing embryo to become a fully formed limb.

The ability to regenerate missing body parts varies according to the complexity of an organism. Generally speaking, simple organisms are better at it than complex ones. Take the simple sponge. You can stick a sponge in a blender, tip the resulting soup into a bowl, and, in a few days, you will have a fully reconstituted sponge. Somehow, the melee of displaced cells is able to reorganize and regroup. The earthworm, a more complex creature, is not so versatile. Even so, decapitation is no more than a minor inconvenience. Salamanders cannot regrow a new head, but they can regenerate missing limbs and tails. As you continue to move up the scale of biological complexity, to birds and mammals, regenerative powers become ever more limited. For humans, replacement hair, nails, and skin is about as much as we can manage.

Because they were cheap and easy to breed, Morgan spent most of his time working with earthworms. This was unfortunate for the earthworms. In one experiment, he cut worms in half and, with a fine needle and thread, deliberately sewed the "wrong" bits back together, creating worms with two heads or two "tails."

Morgan wanted to see what happened when small chunks were cut off the tips of these modified worms. In ordinary circumstances, a normal worm would regenerate whatever missing piece had been severed. But would the same be true in the modified worms?

The experiment proved problematic. Morgan had trouble with the double-headed worms. Probably because his sewing was not up to the job, the two heads were reluctant to stick together, so further manipulations were impossible. He had more luck with the other halves, however: two rear ends seemed content in their unlikely union. Even so, a worm with two tails and no head will not stay alive very long. Clearly, in such circumstances, a head would come in handy. But when Morgan cut off a chunk from the tip of the worm, another tail grew back in its place.

Morgan's experiments may have seemed mad and macabre. But the results gave him some important insights into the limits of regeneration, where the regenerative capacity came from, and how it was controlled. The study typified the experimental approach to biology that was beginning to dominate the mainstream. By the early 1900s, experimental biology was on a roll. The naturalists were finding themselves increasingly marginalized as biology became more and more experimental in outlook.

Morgan was no longer just a participant, he had become an important catalyst in this biological metamorphosis. His attitudes and opinions had become increasingly hard-edged. Grand ideas and theories, he asserted, were worth very little unless they could be backed up with experimental proof.

But it was not all science at Bryn Mawr. In the summer of 1904, Morgan married Lilian Sampson, one of his former graduate students. Their honeymoon was unorthodox, combining summer sightseeing in California with research at Stanford and Berkeley. One can only guess at the topic of bedtime conversation. Sea spiders perhaps? They were certainly not afraid to spray their own gametes around, as four children would later testify.

With the wedding came a decision to leave Bryn Mawr. Despite thirteen happy years, Morgan's tiny college faculty could never provide the intellectual diversity that he yearned for. Besides, a job offer at Columbia University in cosmopolitan New York was too good to turn down.

He arrived at Columbia with a sizable reputation. He was now a zealous advocate of experimental science and a harsh critic of the descriptive diet of his academic youth. He was inquisitive, ambitious, and dedicated to his work. And, at the age of thirty-eight, he was an experimental biologist of world renown.

It was a great time to meet the fruit fly.

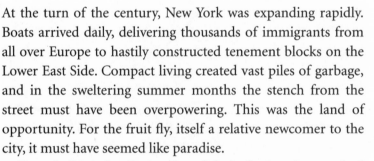

At the turn of the century, New York was expanding rapidly. Boats arrived daily, delivering thousands of immigrants from all over Europe to hastily constructed tenement blocks on the Lower East Side. Compact living created vast piles of garbage, and in the sweltering summer months the stench from the street must have been overpowering. This was the land of opportunity. For the fruit fly, itself a relative newcomer to the city, it must have seemed like paradise.

Years before, the first wave of fruit fly immigrants had arrived at ports in the Caribbean, carried across the Atlantic in slave ships from Africa and southern Europe. In the 1870s, in the immediate aftermath of the Civil War, the burgeoning trade in rum, sugar, bananas, and other fresh fruits delivered them north to Boston, New York, Philadelphia, and other flourishing cities on the east coast.

In the early 1900s, the fly was just one among many animals to land in the laboratory. The explosion in experimental biology had spawned a frenzy of animal prospecting, and few people foraged more widely than Morgan. In his early years at Columbia he studied how sex was determined in aphids, embryonic development in frogs and toads, regeneration in fish, and heredity in wild rats and mice. Morgan arrived relatively late on the fruit fly scene. His first encounter with the fly came in 1907, seven years after the fly's debut at Harvard University.

In that time, the fly had established itself as a reliable, if unremarkable, laboratory workhorse. The insect was seen very much as a stopgap measure, a rough-and-ready option, ideal for student projects, or for use when other, more desirable animals were unavailable. This is the role in which Morgan first cast the fly when it arrived on his doorstep at Columbia.

Morgan had taken on a new graduate student, Fernandus Payne, and needed to set him up with a suitable research project. Payne told Morgan he was interested in the evolution of blindness in cave-dwelling fish, seeing it as a possible example of Lamarckian evolution.

In the early part of the nineteenth century, the French evolutionary biologist Jean-Baptiste Lamarck had argued that organisms evolve in response to their "needs." A change in the environment—moving from the light into the darkness of a cave, for example—would negate the need for eyes. Needs, Lamarck believed, were articulated through the use or disuse of one or more anatomical features. Physical changes brought about by use and disuse were encoded into the sperm and egg and passed on to the next generation—the so-called inheritance of acquired characteristics. It was all very complicated. It was also complete garbage.

Morgan, however, thought otherwise. Like Payne, he was interested in Lamarckian evolution and thought that it might be worth trying to test experimentally in the laboratory. Given the limited time and money available, a laboratory study on some weird and, no doubt, extremely neurotic species of cave fish was out of the question. So, after lengthy discussions, Morgan and Payne plumped for the fruit fly instead.

Payne condemned forty-nine generations of fruit flies to a life of complete darkness. When the last generation of flies emerged, bleary-eyed, into the light of the laboratory, Payne looked to see if their eyes had decreased in size. They hadn't.

To Morgan, the result was less important than the methods. Firsthand experience of rearing the fruit fly in the laboratory had persuaded him that the fly belonged in his own family of research animals. It may have been small and lacking in prestige, but the fly had attributes that made it particularly well suited to the increasing demands of an academic lifestyle.

At the beginning of the twentieth century, Lamarck's name was just one among many to be heard in heated, muddled debates about evolution and heredity. Charles Darwin had come and gone, leaving his footprints all over the conceptual landscape. God had been successfully sidelined, with most biologists now accepting evolution as fact. But although Darwin was much admired as a scientist, not everyone had swallowed his theory whole. The debate was no longer about whether evolution occurred, but how. When he died, in 1882, Darwin left behind a cottage industry of competing theories and a bickering biological community.

Darwin's evolutionary argument was elegantly simple. Animals and plants produce more offspring than their environment can support. This leads to competition between individuals for limited food and living space. Because of small, heritable differences between individuals, some will stand a better chance of survival than others. In every generation, natural selection sifts through the population of competitors, eliminating the unfit and adapting individuals to their environment. It was a powerful argument but, as Darwin himself conceded, it was not watertight.

The Achilles' heel was the lack of any coherent theory of heredity. Variation within a species was unquestionable—you could see it with your own eyes—but how did this variation originate, what was its material basis, and how was it transmitted from one generation to the next? These were questions that would haunt Darwin throughout his later life.

Of course, ignorance was no barrier to speculation. To the mid-nineteenth-century mind, inheritance was mostly a matter of blending. An individual's characteristics were thought to be a mixture or average of those in the two parents. On looks alone, there seemed to be some truth to the idea. A tall father and short mother, for example, invariably produced a child of intermediate height.

There were obvious exceptions to the blending idea. Even the Victorians, who hid their genitals beneath a dozen layers of underwear, knew that nobody inherited half a penis. But cases of either/or inheritance, where a child inherited the features of either the mother or the father, rather than a blend of the two, were considered exceptions to the rule.

When it came to an underlying mechanism to explain blending inheritance, Darwin himself came up with a rather whimsical theory that he called "pangenesis." Darwin imagined that each part of the body produced a miniaturized version of itself—a gemmule—that was transported via the bloodstream to the reproductive organs. Sex would unite gemmules from both parents in the offspring. The gemmules would then multiply to form full-scale versions of the tissues and organs from which they were derived.

Darwin held out great hopes for his pangenesis theory. In 1867, in a letter to the American botanist Asa Gray, he wrote, "The chapter on what I call Pangenesis will be called a mad dream . . . but at the bottom of my mind, I think it contains a great truth." It *was* a mad dream. Ironically, it was Darwin's own cousin, Francis Galton, who sounded the death knell for the pangenesis theory. Galton was, at various times, an explorer, a scientist, an inventor, and a professional racist. Sometimes he was all four things at once. He was also a regular visitor to the fringes of madness. One of his inventions included a top hat with a hinged lid that could be raised by squeezing a rubber bulb. Galton believed that the device provided the necessary ventilation to prevent his active mind from overheating.

More alarming, perhaps, was Galton's obsession with quantifying anything and everything. He would place pressure sensors under the chairs of dinner guests to record their body movements. He once conducted a statistical analysis into the efficacy of prayer, to see how the frequency of prayer affected longevity. (His conclusion, that pious types actually died younger than those less devoted, was strangely reassuring.) He also compiled a beauty map of the British Isles based on his assessment of the number of beautiful, nondescript, and ugly women he saw on the streets of various British towns: "I found London to rank highest for beauty; Aberdeen lowest."

There was more. Much more. He devised a way of assessing the boredom level of an audience by measuring the average rate of fidgeting. He developed complex mathematical formulas for working out the correct amount of tea to drink during the day, based on such crucial considerations as the amount and temperature of the water, and the time taken for brewing. And when he somehow found time to sit for his portrait, he meticulously recorded the number of brush strokes used to paint it. The results were published in a 1905 issue of *Nature* under the title "Number of strokes of the brush in a picture."

So it must have been during one of his more lucid moments that Galton carried out a simple and ingenious experiment to test Darwin's pangenesis theory. Galton transfused the blood of brown rabbits into pure-bred "silver-grey" rabbits. He reasoned that if there was a gemmule for the brown coat color, it would be transferred to the silver-greys with the brown rabbit's blood, and revealed in the next generation. Before long, Galton's house was overrun with silver-grey rabbits.

Pangenesis was history, yet Darwin stubbornly defended his pet theory, protesting that Galton's experiment was inconclusive. For Darwin, public criticism from one of his own relatives, let alone one who was completely bonkers, must have been particularly galling.

With pangenesis disappearing down the plughole, Darwin had other worries to contend with. Several critics had pointed out a fundamental contradiction between blending inheritance and his evolutionary theory. Evolution by natural selection depended on the existence of variation—heritable differences between individuals. But if blending inheritance was the norm, then each successive generation would see a progressive dilution of the variation within a population. Gradually, differences would erode until all individuals ended up the same. If there was no variation, then there could be no origin of species.

The criticism had Darwin cornered. He could not envisage

an alternative to blending inheritance. But neither was he about to ditch natural selection. So to resolve the paradox, he fell back on Lamarck's inheritance of acquired characteristics. If physical changes, brought about by use and disuse, could be cemented into the germ line and passed on to the next generation, then Lamarckian inheritance could be a source of hereditary novelty. Darwin believed that Lamarckian inheritance could counteract blending and maintain the variation that was needed for natural selection to work. If blending implied a continual draining away of variation, then Lamarckian changes would keep topping it back up.

Another of Darwin's critics was St. George Mivart. It was Mivart who, in 1871, made the well-publicized anti-Darwinian attack that has been a mantra of the creationist movement ever since. Mivart argued that complex structures—the eye was the classic example—were adaptive only in their completed form. The workings of an eye depends on a host of interrelated bits and pieces—the lens, retina, muscles, nerves, and so on. Yet Darwinian natural selection insisted that the eye must have evolved gradually, through numerous incipient stages. How, Mivart asked, could these incremental stages have had any adaptive value? An eye was useful, but Mivart could not see how "half" an eye was any better than no eye at all.

Darwin disagreed, arguing that any individual variation, however minor, would be favored by natural selection, provided that it conferred some competitive advantage. Even a primitive eye—a single light-sensitive cell, for example— might enable an animal to detect an approaching predator and give it an edge in the struggle for existence.

Not everyone was convinced by Darwin's argument. Years later, Morgan echoed Mivart's criticism with reference to his own work on regeneration in earthworms. Morgan seriously doubted whether regenerative ability could have evolved gradually through small increments. What use, he asked, would there be in replacing, say, only half a limb? Regenerative ability, Morgan argued, was useful only when a limb could be reconstituted in its entirety. It must, he believed, have arisen all at once, in one single, giant evolutionary leap.

Morgan was not alone in his anti-Darwinian criticism. Many of his peers shared a deep distrust of Darwinism. It was a subject that had become associated with the tired old ways of the naturalists. For the younger breed of experimentalists, there was a new pretender to the evolutionary throne: a Dutch botanist by the name of Hugo de Vries.

De Vries espoused his grand evolutionary vision in *Die Mutationstheorie* (The Mutation Theory), first published between 1901 and 1903. Anyone who has ever driven through the Netherlands and been puzzled by the lack of trees would do well to consider the environmental impact of this encyclopedic work. Containing more detail than was strictly necessary, *Die Mutationstheorie* ran to two shelf-bending volumes.

The seeds of de Vries's mutation theory can be traced to a trip he made to a field site just outside Amsterdam in the early 1890s. While out walking, he came across what looked like three distinct varieties of the evening primrose, *Oenothera*, growing side by side. De Vries was certain that one of the varieties was the parent of the other two. But the differences between the individual plants seemed far greater than the minor variations that Darwin had talked about. In fact, so different were the plants that de Vries believed he was looking at three distinct species, species that had arisen not through the slow and gradual accumulation of minor differences, but instantaneously, in a single large jump. These jumps were what de Vries christened "mutations," although today the word means any genetic change, small or large.

De Vries saw the evening primroses not as sheer serendipity, but as the key to a new evolutionary scheme. He argued that small-scale differences between individuals—the fuel of Darwinian evolution—had nothing to do with the origin of species. To de Vries, new species were the product of giant evolutionary leaps. He did not discount Darwinism entirely. Natural selection would still pick the cream of the crop. But it would pick from a motley assortment of meaty mutants instead of slight, individual variations.

De Vries argued that a new mutant would be reproductively incompatible with its parents. This got around the problem of variation being lost by blending; but it created new problems of its own. In effect, a de Vriesian mutation would create a sexual pariah. Without a sexual partner the new mutant was destined for a lonely life and an evolutionary cul-de-sac. So to counteract this problem, de Vries came up with the idea of a "mutating period"—rapid bursts of mutation during which the same mutation could arise simultaneously in several individuals. According to de Vries, these mutating periods were induced by excessive heat, cold, and other environmental extremes.

The mutation theory was popular because it seemed to circumvent many of the perennial criticisms aimed at Darwinism. It neatly sidestepped the problem of blending, and it addressed Mivart's skepticism about the adaptive value of incipient stages—with de Vriesian mutations there were no incipient stages. But most important of all, for Morgan at least, the mutation theory seemed open to experimental investigation. One of Morgan's long-standing gripes about Darwinism was its curmudgeonly reluctance to succumb to experiments. By proposing that evolution was a slow, long-winded affair, in which perceptible change took place over timescales that far exceeded the human life span, it seemed as if Darwin had deliberately designed his theory so it could not be tested. Of course, this was not an issue to the naturalists, for whom speculation in science was the norm. But to a hard-core experimentalist like Morgan, it was heresy. For Morgan, de Vries's evolutionary ideas seemed to offer the first road sign out of the empirical desert.

Morgan was keen to try to simulate the mutating period in the laboratory. But there were certain experimental obstacles to consider first. Even during mutating periods, mutations were believed to be relatively rare events. So to stand any chance of inducing, let alone finding, a mutant in the laboratory, Morgan would need to look at lots of individuals. He needed something small, cheap, and prolific, a no-nonsense organism whose sole raison d'être was to produce more no-nonsense organisms in superquick time. The fruit fly was ready-made for such a role.

To simulate environmental extremes, Morgan subjected the flies to a barrage of abuse. Flies had acids and alkalis injected into their gonads; they were spun in a centrifuge at eye-popping speeds; and stuck in fridges and ovens for days at a time. But it was all to no avail. The experiments turned out to be a failure for Morgan and torture for the fly. No matter what methods he tried, he could not induce de Vriesian mutations. The mutating period turned into a waiting period.

While Morgan was busy waiting for something to happen to his flies, a new biological craze—Mendel mania—was sweeping across the academic landscape. The man behind the mania, Gregor Mendel, was dead; but his ideas on heredity, exhumed and dusted down, were taking the biological world by storm.

In life, nobody took much notice of Mendel. He died in 1884, an unknown Austrian monk. But it all could have been so different. In 1866, only seven years after the publication of Darwin's *On the Origin of Species,* he had written a modest little paper on breeding experiments in pea plants, in which he outlined some new ideas on heredity.

Mendel was no Billy Graham. A monastic lifestyle and a modest demeanor meant that he was never destined for a life on the world lecture circuit, preaching his new vision of heredity. Even if he had been more outspoken, it is doubtful whether anyone would have listened. Mendel's ideas were way ahead of their time, and completely at odds with the prevailing consensus on heredity.

But in 1900, his hereditary ideas resurfaced, when several biologists finally cottoned on to what he had been talking about in his original 1866 paper. The publicity surrounding the "rediscovery" led many biologists to drop what they were doing and look for similar patterns of inheritance in other species.

Mendel's masterstroke had been to keep things simple. He had restricted his studies to the inheritance of characters that varied in an either/or fashion. Pea plants were either tall or short; peas were either smooth or wrinkled, yellow or green, and so on. After years of study, he began to recognize regular patterns in the way these characteristics were passed on to successive generations. What is more, he was able to interpret these patterns of inheritance in terms of physical phenomena.

The details behind Mendel's ideas deserve a mention, but

too often their explanation descends into textbook tedium. So I'll illustrate them with an analogy. Instead of a row of pea plants in Mendel's monastery garden, think of a row of detached houses in a suburban street, Mendel Street North. It is an unusual street because the characteristics of each house vary, like the characteristics of Mendel's pea plants, in an either/or fashion. The front doors are either black or white, the windows are either square or round, the roofs are flat or sloped, the chimneys are tall or short, and so on.

In this abstract world, each characteristic of a house is determined by an encoded pair of "instructions." The instructions in a pair might be the same or different. Both instructions might say "black door," for example, or both might say "white door." Alternatively, the instructions might be contradictory—one might say "black door," the other "white door"; in these circumstances, the color of the door would be resolved by a kind of intrinsic hierarchy between the instructions. A "dominant" instruction would overrule the instruction of its "recessive" partner. If "black door" was dominant to "white door," for example, then the pairing of a "black" instruction with a "white" instruction would result in a black front door.

Mendel imagined that each instruction was cast in a particle-like form, like a fortune cookie, perhaps. It was a distinct entity that could not blend with its partner and was passed on from one generation to the next, unchanged. The two instructions in a pair would become separated during the formation of the sex cells—the sperm and eggs—so that a sex cell would carry one of the two instructions present in the parent. If the parent had two identical instructions, then all of the sex cells would carry the same instruction. If the parent had two different instructions, then half the sex cells would carry one instruction and half would carry the other instruction. At fertilization, when a sperm and egg fused, one instruction from each parent would combine to form a new partnership.

So imagine if two houses on our suburban street take a fancy to one another and pair up to become semidetached. One of the houses has a black door (with both instructions saying "black door"), the other a white door (with both instructions saying "white door"). The two houses mate and produce a new street of houses—Mendel Street South. What color will the front doors be? All the houses on Mendel Street South will have inherited a "black" instruction from one parent and a "white" instruction from the other. Because black is "dominant" to white, all the front doors will be black. The white door characteristic (but not the instruction) will have disappeared.

But what would happen if two of these new houses found romance and produced yet another street of young houses—Mendel Street East? Clearly, environmentalists would be unhappy about the pace of urban development. But, more importantly, how would the front doors look?

Every house on Mendel Street South carries both a "black door" instruction and a "white door" instruction. Determining which of these instructions a new house on Mendel Street East inherits from each parent is like tossing a coin and seeing if it comes up heads or tails. So taking Mendel Street East as a whole, you would expect, on average, a quarter of the houses to inherit two "white" instructions, a quarter to inherit two "black" instructions, and half to receive one "black" and one "white" instruction. In other words, Mendel Street East would have three black-doored houses to every white-doored house. This three-to-one ratio became the signature of Mendelian inheritance.

When they first appeared, in 1866, Mendel's ideas probably did seem about as relevant to the real world as two houses having sex. Although his hereditary scheme was all very neat and tidy, there was no direct physical evidence to back it up. In the 1860s, the microscopic world was still shrouded in mystery. The technology was not available to look inside a cell, let alone to catch a glimpse of one of Mendel's hypothetical particles.

But by the 1880s, the mist of ignorance was beginning to lift. With advances in microscope design, and the application of selective chemical stains and dyes borrowed from a burgeoning textiles industry, the image of the cell was slowly transformed. A transparent, featureless desert gave way to a landscape of candy-colored shapes and contours. The cell had an internal anatomy all its own, complete with microscopic renditions of organs and tissues.

The long lens of the microscope connected two quite different worlds, one human and familiar, the other alien and microscopic. To look down the eyepiece was to take in a biological peepshow, to enter the surreal and private realm of cellular theater. And center stage, deep in the heart of the cell, was a cast of deeply stained, wormlike structures. Biologists called them chromosomes.

By the turn of the century, many biologists were touting chromosomes as likely candidates for the bearers of the hereditary material. Chromosomes appeared to give real physical substance to Mendel's hypothetical particles. They came in pairs, with the mother and father making equal contributions to both. And each chromosome in a pair separated from its partner during the formation of sperm or eggs.

The coincidence was certainly not lost on Walter Sutton, a graduate student at Columbia University. In 1902 he wrote:

> I may finally call attention to the probability that the association of paternal and maternal chromosomes in pairs and their subsequent separation . . . may constitute the physical basis of the Mendelian law of heredity.

Perhaps fearful that his career could only go downhill after such an important insight, he immediately gave up biological research and became a surgeon.

Despite the lack of hard evidence, there was a definite momentum moving biology in the direction of chromosomes and Mendel. The language of inheritance was changing to accommodate the finer points of Mendel's theory. In 1909, Mendel's particles were renamed "genes" by Danish biologist Wilhelm Johannsen, and their study became genetics.

Morgan, meanwhile, was in denial. In his early years at Columbia, he had routinely rubbished Mendelian inheritance and the chromosomal basis of heredity. His initial reticence toward Mendelian ideas was entirely predictable. In Morgan's view, Mendel's hereditary scheme was all symbols and theorizing, an abstract invention with little factual basis in reality. It was everything he liked least in science.

What is more, Morgan could not accept that the simple either/or characteristics that Mendel had studied bore much resemblance to variation in nature, where the evidence suggested much more complex relationships. To Morgan, the Mendelian housing estate was a land of make-believe. In the real world, front doors were not just black or white, they were also red, blue, green, yellow, and all colors in between. In 1909, he summed up his feelings to a largely pro-Mendelian audience at the American Breeders Association in St. Louis:

> ... I realize how valuable it has been to us to be able to marshal our results under a few simple assumptions, yet I cannot but fear that we are rapidly developing a sort of Mendelian ritual by which to explain the extraordinary facts of alternative inheritance.

When it came to the alleged involvement of chromosomes in heredity, Morgan was again the consummate scientific party pooper, keen to play up any evidence to the contrary. In the early 1900s, the German biologist Theodor Boveri had come up with two powerful pieces of evidence in support of the chromosome theory. In one experiment, Boveri demonstrated that a complete set of chromosomes was essential for the normal development of a sea urchin embryo. And in another, he was able to show that the chromosomes of the roundworm, *Ascaris,* have a physical integrity that persists from one generation of cells to the next.

Observations of dividing cells in other species, however, gave the impression that the chromosomes performed their very own Houdini trick. During cell division, the chromosomes appeared as bright as day. But when the cells had stopped dividing, they seemed to dissolve away, only to miraculously reform, like clouds in the sky, when it was time for the cell to divide again. In Morgan's view, chromosomes were too mercurial to have any substantive link with heredity. In 1906, he wrote to his friend Hans Driesch, "I am glad you are going to examine Boveri's experiment. I have always distrusted it, but until it is cleared up, the chromosomal people will find it convincing."

But more evidence was emerging in support of Sutton's

prophetic judgment on the chromosomal basis of heredity. Several biologists, for instance, had detected a startling and consistent difference between the chromosomes of males and females, a difference that mapped to a single chromosome pair.

In the early 1890s, biologists were muttering about a chromosome that seemed to lack a partner. Because of its enigmatic nature, this chromosome loner was designated the "X" chromosome. But years later, a partner for the X chromosome did materialize in the form of a much smaller, stubby little thing, dubbed the "Y" chromosome. Only males, it seemed, had this ill-matched pairing of an X and a Y chromosome. Females had a more balanced partnership, made up of two same-shaped X chromosomes. This pattern in the so-called sex chromosomes—males XY and females XX—was seen first in beetles, and later confirmed in grasshoppers, flies, and many other animals. It seemed to be the first direct evidence linking inheritance—in this case the inheritance of sex—to chromosomes.

The evidence seemed watertight until the chromosomes of birds and butterflies came under the microscopic spotlight. In these animals, the difference between males and females was the exact opposite of that seen so far: it was the females, not the males, whose chromosomes were ill-matched. Sex determination was turning out to be more complicated than expected.

What was needed was a hereditary visionary to shake some sense out of the confusion, a genetic saint who could lead the muddled masses toward the promised land. On the face of it, Morgan was an unlikely candidate for this demanding role. In fact, it is hard to imagine anyone less likely to take up the challenge. But Morgan was nothing if not independent. His scientific opinions could change with the seasons. All that mattered was the strength of the experimental evidence in front of him.

One winter's day in early 1910, Morgan was routinely perusing his flies. His stocks had grown—he was now maintaining a sizable collection—but he had no reason to suspect that this particular day was going to be different from any other. There may even have been a mood of disenchantment in the air. Morgan had more or less given up hope of ever finding one of de Vries's elusive mutations and was moving on to other things.

But on this day he spotted something unusual in one of the bottles. Keen to take a closer look, he anesthetized the flies with a heavy whiff of ether before tipping the bottle's contents onto his desk. Picking through the dozy and delicate bodies, he separated the odd-looking fly from the rest.

Morgan reached into his jacket pocket for his magnifying lens and began to scrutinize the tiny insect in front of him. The fly was male—the heavy splodge of melanin on the tip of the abdomen told him that. But it was the head that caught his attention. Bringing it sharply into focus, two white, expressionless eyes leaped up at him through the thick glass of the lens.

Morgan had only ever known flies with red eyes. This white-eyed fly was obviously a new mutant. But it was hardly the kind of mutant envisaged by de Vries. Apart from the change in eye color, the fly was evidently the same species as all the other flies. The mutation looked like the kind of simple, small-scale change that Darwin had talked about.

Morgan decided to try to breed his white-eyed male with a normal red-eyed female. To his relief, the two flies got on famously. Eye color was no barrier to sexual attraction. The next day, Morgan watched the mated female carefully deposit her fertilized eggs in the yeast-rich substrate he had prepared for her. Within hours, the eggs had hatched and a writhing mass of miniature maggots was tucking into the feast.

Morgan's only interest was in how the eyes of the adult flies would turn out. But he had to be patient. There was a week to wait while the blind, featureless grubs went about their business, and then another week while the maggot body was remolded and reshaped into its adult form.

Finally, after a frustratingly long wait, the first adult began to emerge. A fresh-faced fly pierced its pupal case and looked up into the light. Morgan was there, waiting expectantly, straining for an early glimpse of the eyes. They were a brilliant red.

Within seconds, another fly had slipped out. It, too, had normal red eyes. So did the next. And the next. One by one, red-eyed flies emerged until all the pupal cases were empty. Every fly had normal red eyes. The white-eye characteristic had disappeared. It was just as Mendel would have predicted, if the instruction for red eyes was dominant to the one for white eyes. So Morgan took the experiment one stage further, as Mendel would have done, and paired up brothers and sisters from this new generation of red-eyed flies. It was incest by any other name, but ever the opportunists, and completely bereft of scruples, the fruit flies seemed only too happy to oblige.

It was another frustrating wait before the next generation of adult flies emerged. But as the first few flies started to struggle their way out into the world, one thing was obvious. This time, the flies were not all the same. Some of the flies had red eyes and some had white eyes.

White eyes had disappeared in one generation, only to

reappear in the next—again, just as Mendel would have predicted. But what was the ratio of the two different types of fly? Did this also conform to Mendelian expectations? Morgan carefully sorted through the thousands of flies, counting the numbers of each type. There were 3,470 flies with red eyes and 782 with white eyes. Within the limits of probability, the figures were near enough. Here was Mendel's notorious three-to-one ratio, right under Morgan's nose.

Morgan noticed another striking pattern in this generation of flies. Although males and females were produced in roughly equal numbers, there was a very unequal distribution of eye colors between the sexes. There were 2,459 red-eyed females, 1,011 red-eyed males, 782 white-eyed males, but no white-eyed females. Here was something that Mendel could not have predicted: the white-eye characteristic had been transmitted exclusively to the grandsons. In later breeding experiments, Morgan would show that white eyes were not limited to one sex alone. In certain types of crosses, females could inherit white eyes. But the trait was always much more common in males than in females.

Instances of genetic characteristics being associated with one sex more than the other were not new. They had already been recorded in birds and butterflies. But in these animals, the bias was in the opposite direction: it was the females, not the males, who were more commonly affected. Now Morgan had shown the opposite pattern—a link with males.

To his credit, Morgan put his opposition to Mendel and chromosomes to one side as he sought an explanation for these apparently contradictory observations. He began to realize that the observations made sense if you thought in terms of the sex chromosomes rather than the sex of the individual. After all, the one thing that female birds and butterflies have in common with male fruit flies is a single X chromosome.

Now what if a gene was carried on the X chromosome? What kind of outcome would you expect? Morgan was beginning to think thoughts he had routinely derided in others. But he could not help himself. The fly had forced him to free his mind of prejudice and rethink all rational solutions. Armed with his thought experiment, he allowed his mind to wander wherever the logic led him.

Male flies carry a single X chromosome, which they inherit from their mother, and a Y chromosome, which they inherit from their father. Female flies carry two X chromosomes—one from the mother and one from the father. Imagine that the gene for eye color is on the X chromosome. Males would inherit only one genetic instruction for eye color (the Y chromosome is too puny to carry any corresponding instructions), whereas females would inherit two instructions.

Because the white-eye instruction is recessive to the red-eye instruction, a female fly would need two copies of the white-eye instruction—one from her mother and one from her father—to inherit white eyes. But a male fly would need only a single white-eye instruction from his mother to inherit white eyes.

The chances of a male inheriting one copy of the instruction would be greater than the chances of a female inheriting two. It's a bit like tossing a coin twice. The chance of getting at least a single head—the white-eye instruction—in two throws is much greater than getting two heads. In other words, you would expect white eyes to be more common in fruit fly males.

The logic seemed watertight. If genes are carried on the X chromosome, then any recessive characteristic would be more common in the sex with the single X. The reasoning held true for fruit flies, birds, butterflies, even ourselves. In the nineteenth century, for example, many of Queen Victoria's descendants suffered from the hereditary disease hemophilia, but it was men who were disproportionately affected. Morgan noticed that the inheritance of red-green color blindness in humans follows exactly the same pattern of inheritance. Red-green color blindness is much more common in men than in women.

Morgan had conjured up a compelling amalgam of hereditary ideas. He had unified genes, chromosomes, and sex determination into a single coherent story, a story that offered enormous explanatory power. The white-eyed fly had induced a remarkable transformation in his scientific outlook. It had eroded his opposition to all things Mendelian and chromosomal, leaving him with a new material vision of biology. Life, for both Morgan and the fly, would never be the same again.

The white-eyed mutant, or *white** for short, was just one of many new fly mutants to materialize that year. Between June

---

*Ever since Morgan, newly discovered genes have been named after the mutant by which they were first identified. So *white*, for example, refers to an eye-color gene and not just the instruction for white eyes. To distinguish the different versions (alleles) of the gene, various suffixes are used. A white-eye instruction might be written as *white*⁻, for example, and a red-eye instruction would be *white*⁺. The first letter of the name can either be capitalized or lowercased, depending on whether the mutant form of the gene is dominant or recessive to the normal or "wild-type" version. In this instance, *white* is written as lowercase, because the white-eye allele is recessive to the red-eye allele.

Somewhat confusingly, the gene name can also be used in a different context—as it is here—to identify the mutant fly. All flies carry some version of the *white* eye-color gene, irrespective of their eye color, but only white-eyed mutants would be identified as *white* flies.

and August 1910, Morgan discovered a trio of wing mutants: *rudimentary, truncate, and miniature*—flies with normal-sized bodies but tiny, stunted wings. Then there was *olive*, a fly with an olive body color instead of the normal allover tan, and *pink*, another eye mutant. All these genetic novelties were recessive variations on the normal theme, and all displayed Mendelian patterns of inheritance.

There was nothing mysterious about this flood of new mutants; it all boiled down to a change of scale. Having abandoned his search for de Vriesian mutations, Morgan had decided to scale up his fly operations in preparation for a new and unrelated study on experimental evolution. Instead of a few hundred, he was now rearing tens of thousands of flies at a time.

Mutations are rare events. You can think of a mutation as being a bit like a winning lottery number, and the flies being the lottery tickets. If you only buy a few tickets (or keep a few flies in your laboratory), then there is only a very slim chance that you will win (or see a new mutation). But increase the number of tickets (or flies) to thousands or tens of thousands, and the odds start to fall.

More flies meant more crosses, and more crosses further boosted the chances of seeing a new mutation. Many mutant alleles are recessive instructions, so they will be concealed by their dominant partner when they first appear. The only way of discovering the new instruction—of seeing its effects—is to get two flies, both are whom are carrying the same instruction, to mate. Of course, initially, this is a hit-or-miss affair, because you don't know which fly is carrying which instruction. But increase the number of crosses and you increase the chance of two of the same recessive instructions being united.

Once Morgan had identified a mutant, he would make

controlled crosses to obtain a stock of male and female flies that carried only the new mutant version of the gene. These crosses further exposed more mutants, which, in turn, led to more crosses. In a matter of months, Morgan had turned his laboratory into the fruit fly equivalent of a nuclear reactor.

Morgan was enthralled by the fly. In November 1910, he wrote to his friend Hans Driesch:

> It is wonderful material. They breed all year round and give a new generation every twelve days.

But Morgan was rapidly becoming a victim of his own scientific success. The work needed to maintain the ever-growing crop of new mutants was overwhelming him, and the fruit fly reactor was in serious danger of meltdown. In March 1911, he wrote:

> I am beginning to realize that I should have prepared for a large campaign and be better organized, but who could have foreseen such a deluge. With vicarious help I have passed one acute stage only I fear to pass on to another. With what help I can muster I hope to weather the storm.

At the end of 1910, help had arrived in the form of two keen undergraduate students, Calvin Bridges and Alfred Sturtevant. Both were given desks in room 613, later nicknamed the "Fly Room," next door to Morgan's office, on the top floor of Columbia University's Schermerhorn Hall.

In today's health-and-safety-obsessed times, the Fly Room would probably be closed down immediately. Even then, the room was considered audacious in its approach to cleanliness. Visitors who arrived expecting some kind of pristine monument to the experimental method were taken aback by the mess and the dirt. If the idea was to make the fly feel at home by re-creating the atmosphere of its natural environment—the dustbin—then the Fly Room did the trick.

The room itself was small, less than five by seven meters. Eight wooden desks, overloaded with trays of half-pint milk bottles and microscopes, covered the floor. More bottles spilled over onto disordered sets of shelves. Sketches, maps, charts, and memos papered the walls. In one corner sat a dubious-looking sink piled high with steel pans and ladles, all bearing the stains and bruises of frequent use. Behind the sink, a blackening bunch of bananas hung from the wall. The air in the room was thick with a heady mix of rotting fruit, yeast, and the sickly sweet smell of ether.

During the freezing New York winters, the Fly Room became the focus of Morgan's world, the place where he and his research staff would plan, debate, and carry out their scientific campaigns. In the summer, the room was dismantled, packed into barrels, and shipped off to the Marine Biological Laboratory at Woods Hole on the Massachusetts coast. There, fly operations would continue in the more relaxed seaside surroundings. A few bottles of flies were always left behind as insurance, just in case something unpleasant happened to those in transit.

The mutants continued to flood in. At least ten appeared in 1911. Double that number turned up the following year. By 1914, the total number of mutants had risen to over a hundred. The white-eyed fly may have sown the seeds of Morgan's hereditary ideas, but as new mutants continued to materialize, Morgan was able to refine his hereditary vision and create a more detailed picture of how genes and chromosomes fit together.

Nobody knew how many genes the average plant or animal contained, but the consensus was that it was far more than the number of chromosomes. The fruit fly, for example, has only four pairs of chromosomes. Admittedly, a fly is not one of evolution's most advanced offerings, but no one was seriously suggesting that it was designed around only four pairs of genetic instructions.

If there were more genes than chromosomes, then the independence of many genes must be thwarted by virtue of being physically linked together on the same chromosome, like convicts in a chain gang. Genes that were linked together must be inherited together. This was the theory, at least. The reality was annoyingly different. Examples of two or more characteristics always being inherited together were surprisingly rare.

By the summer of 1910, Morgan had discovered two genes, *rudimentary* and *white*, which seemed to be linked together on the same chromosome. Rudimentary wings, like white eyes, were much more common in males than in females, suggesting that *rudimentary*, like *white*, was on the X chromosome.

If the links connecting genes on a chromosome were unbreakable, then white eyes and rudimentary wings would always be inherited together. But Morgan found nothing of the sort. He was able to breed plenty of red-eyed flies with rudimentary wings and white-eyed flies with normal wings. In fact, the two genetic characteristics behaved completely independently, as if they were situated on different chromosomes. Obviously, the links were not unbreakable.

Morgan believed that chromosomes must occasionally break to allow genetic instructions to be exchanged between partner chromosomes. He didn't just pluck this idea out of thin air. It had already been suggested by the Belgian chromosome wizard Frans Janssens in 1909. Janssens had gained a unique glimpse into the way in which partner chromosomes in a pair behave before being separated and packaged off into the sex cells. One moment the two chromosomes were vertically aligned in parallel, the next they were entwined around one another like two amorous snakes. Janssens proposed that during these moments of physical intimacy, the chromosomes break at corresponding points along their length and exchange complementary segments.

From his own microscopic observations, Janssens could not tell whether the chromosomes actually exchanged bits of themselves or not; all he could make out was their clasping embrace. But Morgan's studies with *white* and *rudimentary* provided genetic evidence to support the idea that these exchanges really did take place.

Not all genetic links were as easily broken as those between *white* and *rudimentary*. With the discovery of more and more X-linked mutants, Morgan found varying degrees of association between linked pairs of genes, ranging from a complete association to no association at all.

Morgan believed that the simplest way to explain varying strengths of association was to start by assuming that genes were linearly arranged on a chromosome, like beads on a string. Each gene occupied a specific location on the chromosome, which corresponded with the position of its partner on the opposite chromosome.

The shuffling of genes between the paired chromosomes was analogous to the shuffling of two packs of cards. The closer two cards are to one another in a pack, the less chance they have of being separated by shuffling. Likewise, the degree of association between any two linked genes depended on their physical proximity to one another on the chromosome.

Sturtevant immediately picked up on the wider implications of Morgan's logic. He realized that you could use the degree of association between pairs of linked genes to work out the linear order and relative spacing of genes along a chromosome. You didn't need any fancy gadgets or gizmos. All you needed were flies—with their inborn and insatiable appetite for sex—and the ability to count. In 1911, Sturtevant produced the first-ever genetic map, a simple affair showing the linear arrangement of five X-linked genes.

Genetic mapmaking was a major step forward. It meant that each new mutant gene could be assigned a position on a chromosome relative to its neighbors. More importantly, perhaps, maps gave genes and chromosomes a visual quality they had hitherto been lacking. A chromosome was now like a stretch of railway line, with genes marking the relative positions of stations along its length.

Much of the work within the Fly Room was concerned with mapping the never-ending supply of new mutants. By 1915, maps had been made for each of the four fruit fly chromosomes, showing the relative positions of the hundred genes discovered so far.

The heavy workload was lessened a little in 1912 when Morgan, Sturtevant, and Bridges were joined by Hermann Muller, a young postgraduate student. Together, they formed a formidable quartet of egos and intellectual ambitions. Reminiscing about these times, Sturtevant wrote:

> There was an atmosphere of excitement in the laboratory, and a great deal of discussion and argument about each new result as the work rapidly developed.

But behind this bland pronouncement on the supposed bonhomie of the cooperative scientific experience lurked an inevitable tension between individual members of the group.

Sturtevant, Bridges, and, particularly, Muller seemed to resent Morgan for taking undue credit for what were often collective discoveries. Commenting on Morgan's part in the fly group's success, one Columbia University colleague is alleged to have remarked: "Morgan has made one and only one important discovery in his life; that discovery was Sturtevant."

Whatever the relative merits and contributions of each member of the group, the overall success radiating out from the Fly Room was not lost on the outside world. Morgan, his students, and the fly had come as close as anyone to a foolproof explanation linking Mendelian inheritance with the chromosome theory of heredity. What is more, they had turned fruit fly breeding into a mapmaking art. Together, they had assumed the role of pioneers in the new and fledgling field of genetics.

Morgan's success was also a great advertisement for the fly. Old Victorian favorites such as rats and mice were rapidly displaced as the fly assumed the mantle of laboratory superstar. In turning Morgan's biological opinions upside down, in coercing him into thinking in terms of genes and chromosomes, the fly had consigned itself to a new, domesticated life indoors.

2

UNSCRAMBLING THE EGG

I knew a man once who had an eye on the end of his big toe. The toe itself was not in its orthodox position. It was stuck to his face where the nose should have been. And the nose? That was inside his abdomen, somewhere between the liver and the spleen. He did, at least, have a conventional set of two, well-matched legs. But they were attached to his eye sockets, and protruded from the top of his head like a pair of antlers. I didn't know this unfortunate man long. Just long enough for his features to imprint themselves on the memory before he dissolved from my dream.

What was on my mind to make this muddled form of a man appear in my sleep? Was I taking my own body for granted? Was I concerned about the shape of my nose? The matter remained unresolved.

But years later, I came to see the dream from an entirely different perspective, one born, perhaps, of my biological training. These kinds of bodily deformities may seem like figments of a warped imagination. But in fruit flies, they are real.

Back in the late 1970s, extreme mutants were all the rage.

Take a trip round a hip and happening fruit fly laboratory and you might have been forgiven for thinking that you had stumbled across a fruit fly house of horrors. In the search for new mutants, flies were being force-fed mutagenic chemicals and leaving a trail of disfigured descendants in their wake. It was mutants on demand, and there was no limit to the deformity on display.

Take *bicaudal,* for instance, a mutant embryo born without a head, or much of a body. What it did have, however, were anuses. Two of them, in fact, fused back-to-back. Lacking brain, eyes, and any form of locomotory appendages, *bicaudal* had little choice but to arse about for the two or three hours of its short life on Earth.

At least *bicaudal* fared better than *sieve.* Poor *sieve* seemed to have taken The Who's iconoclastic sixties anthem "My Generation" all too literally. It would arrive in the world as a fertilized egg and depart, not many minutes later, with barely enough time for its body to develop the most rudimentary bits and pieces. "Hope I die before I get old," indeed.

The mutants *patch, runt,* and *hunchback* were a motley trio in the first flush of their embryonic youth. Sadly, this was about as far as their lives would take them. Despite having recognizably maggoty features, closer inspection would reveal that various bits of their bodies had gone missing.

But not all the mutants were stuck at developmental base camp. There was *Antennapedia*, for instance, or *Antp* for short. *Antp* was a mutant with an excess in the leg department. But the extra pair of legs were not where you might expect them. *Antp* had two legs sticking out of its face where the antennae were supposed to be.

There were dozens of others, bearing bizarre names like

bazooka and *Bubble, spook, popeye, gooseberry,* and *bladder-wing.* All these mutants suffered from some kind of wholesale rearrangement of their body plan. Some of the deformities were so extreme that the incipient flies could not even make it out of their egg case, and life spans were measured in minutes and hours rather than days.

If these flies had been looking for someone to blame for their wretched predicament, then Hermann Muller would have been a good place to start. To many people Muller was a maverick and a visionary. To the fly, however, he was a cruel tyrant. It was Muller, after all, who pioneered the artificial induction of mutations, paving the way for the mutant blooms of the 1970s.

Mutations have always been the first foot in the door to any genetic study. The only way of finding out what genes do in their normal state is to look at what happens to an organism when things go wrong—when genes mutate. Just as a car mechanic studies the symptoms of a faulty engine to locate the part that is affected, geneticists use the symptoms of mutants to identify genes and their functions.

In the early years of fruit fly research, nobody knew how to induce mutations artificially, so biologists had to wait for new ones to occur naturally. This was a most unsatisfactory state of affairs. It was like waiting for a London bus. Nothing might happen for ages and then two or three might come along all at once.

Attempts to generate genetic damage by artificial means had been tried before. In the first decade of the twentieth century, Morgan had subjected flies to a barrage of biological abuse in his search for de Vriesian mutations. But with no clear understanding of the biological basis of mutation, or the genetic techniques required to detect them, the efforts of Morgan and others had ended in disappointment. Fifteen years later, and with a more refined understanding of the gene, Hermann Muller was to have more luck.

Muller, a onetime member of Morgan's inner circle, had left Columbia University in 1920 to join the University of Texas at Austin. Recognizing the urgent need to find ways of inducing mutations, he had begun to experiment with the effects of heat, but soon switched his attention to X rays.

In 1926, Muller discovered that X rays could cause a massive increase in the mutation rate. Although the exposed flies suffered no obvious visible changes, their genes came in for a pounding. Only in subsequent generations did the true extent of the genetic damage reveal itself. Heavy doses of X rays had caused the mutation rate to rise by *15,000* percent.

Many of the mutants were instantly recognizable. Flies with white eyes, miniature wings, and forked bristles, for example, were the same mutants that had turned up years earlier in Morgan's Fly Room. Muller came to the conclusion that the genetic changes produced by X rays must be identical or similar to those that arise spontaneously.

Muller also found that X rays could cause chromosomes to tear and break. Sometimes, sections of chromosome were inverted so that the linear order of genes was reversed. At other times, stretches of chromosome were deleted, or moved to an entirely new location within a chromosome. These kinds of mutational changes could spell grim news for any fly that inherited them.

Muller's experiments provided the first clear-cut demonstration of artificially induced mutations, and led him on to a lifelong interest in the biological effects of radiation. After seeing what X rays could do to the chromosomes of fruit flies, he made the inevitable link and began raising public awareness of the dangers of radiation to human health.

Muller feared that the increased industrialization of society was bringing with it an increase in the level of environmental radiation and other mutagens. This, he believed, could place an unprecedented burden on the genetic health of the human race. In fact, he became so convinced that human populations were heading for mutation overload that he suggested that brainy men should have their sperm preserved and stored for use by future generations, before things got too bad.

But his prognoses were overly pessimistic. Although he was correct in identifying the hazards of environmental mutagens, his apocalyptic predictions for the human race were partly undone by future discoveries, which showed that cells were not defenseless against the environmental onslaught. We now know that cells have their own biochemical tool kits, which can locate and repair genetic damage caused by radiation and other mutagens. Despite this, Muller remains something of a folk hero for the antinuclear movement.

Muller was a deeply serious man. Even as a child, he seems to have adopted the earnest approach. While most of his contemporaries were happy kicking a ball around the park, Muller was busy laying the groundwork for his own social and scientific philosophy. The seeds of his vision can be found in his unpublished autobiographical notes:

When I was about eight years old, my father took me to the American Museum of Natural History, and . . . made clear to me, through the simple example of the succession of fossil horses' feet shown there, how organs and organisms become gradually changed through the interaction of accidental variation and natural selection. . . . And from that time the idea never left the back of my head, that if this could happen in nature, men should eventually be able to control the process . . . so as greatly to improve on their own natures.

Muller espoused a kind of socialist eugenics. He wanted societies to take control of their own biological evolution, and advocated greater cooperativeness and higher intelligence as worthwhile evolutionary goals. Individual self-interest, he argued, should be subordinated to the social and biological good.

It is intriguing—some might say suspicious—that he chose intelligence rather than, for example, physical beauty as one of his eugenic aims. Muller himself was short, portly, and bald. He also needed glasses to correct his defective vision. But he must have considered himself good breeding stock because he fathered two children of his own.

His left-wing sympathies, however, did little to endear him to the United States government and, in 1932, amid growing personal pressures, he waved good-bye to the Stars and Stripes. A failed marriage, a nervous breakdown, and a homeland hostile to his politics convinced him that it was time for a change.

After a year at the Max Planck Institute in Berlin, Muller moved to the Institute of Genetics in Moscow, at the invitation of Nikolai Vavilov, one of the Soviet Union's foremost Mendelian geneticists. Soviet Russia was more in tune with Muller's socialist sympathies, and Muller obviously saw it as the perfect environment in which to develop his scientific and eugenic ideas.

But his arrival coincided with the start of political and scientific upheaval in the Soviet Union. By 1933, Stalin's paranoia was beginning to translate into a reign of terror. Nobody was safe from suspicion, and that included geneticists. Stalin had a profound distrust of Mendelian genetics. He despised the idea that pea shapes, let alone human characteristics, were determined by genes. Lamarckian inheritance, which emphasized the importance of the environment in shaping the individual and society, sat much more comfortably alongside Marxist philosophy.

Stalin found an ally in the biologist and political opportunist Trofim Lysenko. Stalin's decision to appoint Lysenko as his director of agriculture would effectively put Soviet genetics on hold for the next thirty years, and bring Soviet agriculture to its knees. Mendelian genetics was dismissed as a bourgeois capitalist conspiracy against Marxism, and geneticists were given a stark choice: renounce their allegiance to Mendel, or accept a one-way ticket on the Vladivostok Express.

In 1937, amid a climate of growing unease, Muller decided to get out of the country while he still had the chance. His friend and colleague Nikolai Vavilov was not so fortunate. After being interrogated for over 1,700 hours, a five-minute trial found him guilty of crimes against the state. He died in a prison camp in 1943.

On his way back to the United States, Muller stopped off in Spain for a brief stint opposing Franco's forces in the Civil War. He finally returned to the States in 1940. Six years later, his work on X-ray-induced mutations finally gained formal recognition when he was awarded a Nobel Prize.

Muller's experiments with X rays exposed the dangers of working with radiation and prompted a search for new ways of inducing mutations. After the Second World War, chemical mutagens were incorporated into the fruit fly diet, and replaced X rays as a safer method of generating mutations.

Safer, that is, for the scientist. For the fruit fly, it meant more of the same misery.

Muller had ushered in the era of the instant mutant. But in doing so he had turned the fruit fly's life upside down. The laboratory suddenly lost its laid-back atmosphere. In milk bottles throughout the land, an ambience of comfort and complacency was displaced by one of apprehension and panic. From now on, flies would live in constant fear of the unexpected afternoon snack, the nasty taste in the mouth, and the worry that their offspring might have a head where the anus was supposed to be.

Life for these mutant flies may have been wretched and, in some cases, virtually nonexistent. But deformity has never been so informative. Flies died in their thousands so that we might understand one of the greatest puzzles in biology. A production line of mutant flies was all part of one giant effort to unravel the mystery of embryonic development.

We all start life as a fertilized egg, a single cell tumbling down the oviduct of our mother. That cell divides into two cells, two cells become four cells, four become eight, and so on. Each round of cell division brings with it a doubling of the number of cells. As the embryo grows, cells begin to take on various roles, to form the many different tissues of the body—blood, skin, nerves, bone, muscle, and so on. Some cells even commit suicide, dissolving away to bring limbs, fingers, and toes into sharp relief. Slowly, the recognizable form of a human being comes into view.

But how does each one of the billions of cells "know"

whether it should become a muscle cell, or a nerve cell, or any of the other cell types? After all, every cell carries an identical genetic recipe, the same mixture of genetic instructions that came together when the father's sperm fused with the mother's egg. What makes the skin develop on the outside of the body, rather than the inside? What directs the heart to develop within the body cavity, instead of between our legs? And what ensures that two eyes grow on the front of our face rather than on the end of our big toes? In short, what prevents us from ending up like the man in my dream?

To put things into a broader perspective, think of a building site instead of a body. When a new house is being built, an architect oversees the design, planning, and organization to ensure that everything ends up in the right place. Without the architect, the front door could end up on the second floor, the roof could find its way into the foundation, and a bathroom en suite could turn into a bathroom alfresco.

For decades, biologists have struggled to identify the architects of the body's building site. How does a fertilized egg become a fully grown organism with all its bits and pieces in the right places? The question has been a perennial head-scratcher for embryologists. Take Thomas Hunt Morgan. As his own career in embryology developed, his hairline receded like the ice in spring (which possibly accounts for his attempt to establish some compensatory growth on his chin).

Long before he knew anything about the precise nature of genes, Morgan had struggled with the problem of how cells take on different identities during development. In the 1890s, he had pursued the question through his work on regeneration rather than embryonic development, but the principles, he believed, were the same. In both cases, cells became specialized in a precise and ordered sequence. But what was telling each cell which role it should pursue?

In one of his earthworm-decapitation experiments, Morgan noticed that the farther back he made the cut, the longer it took for the head to regenerate. Heads grew back much more quickly if the cut was made around the "neck" than if it was made farther down the body. Morgan suggested that some kind of chemical gradient inside the worm might be responsible for the varied response. In 1897 he wrote:

> . . . we might speak of the cells of the worm as containing a sort of stuff that is more or less abundant in different parts of the body.

The capacity to regenerate, he argued, depended on the local concentration of stuff, which varied along the body axis of the worm. Recognizing that his explanation was no more than speculation, something he routinely derided in others, he followed it with a typically honest confession:

> I do not pretend that this explains anything at all, but the statement covers the results as they stand.

Speculation or not, Morgan's insight was compelling, and his work prompted a flurry of interest in the study of chemical gradients.

To understand the basic gradient idea, think of a climatic gradient on a mountainside and the way it affects the kinds of plants that can grow along it. The higher up the mountainside you go, the colder it gets. At any point along the slope, the local conditions select only a subset of the many plant species available.

Chemical gradients were thought to work in a similar way. A chemical might, for example, increase in concentration from the tail end of an organism to its head. At any point along the gradient, the local concentration of the chemical somehow directed cells to take on roles that were appropriate to their location within the body. High concentrations at the head end, for example, might instruct cells to develop eyes and brain; low concentrations, a digestive system and genitals.

For sixty years, the study of chemical gradients attracted a small but dedicated following. But the subject was rarely free from criticism. One of the main problems with the gradient idea was that nobody had a clue what any of the chemical substances might be. By the 1960s and 1970s, patience was wearing thin, and dissenting voices were dismissing the whole subject as fuzzy and imprecise. Gradients were falling out of favor, as scientific attention turned more and more toward genes.

Thirty years after Morgan had pinpointed the location of genes, there was still no clear consensus on what genes were made of or how they worked. By the 1940s, many biologists were heralding deoxyribonucleic acid (DNA) as the genetic material. But there were still some dissenters who doubted whether the molecule had the necessary complexity to carry hereditary information. It was not until the early 1950s, when Watson and Crick published the double-helix structure of the DNA molecule, that any lingering doubts over its role were erased.

Following Watson and Crick's discovery, knowledge about genes became progressively more refined. The DNA molecule turned out to be a code that contains information in the sequence of its four chemical letters, A, G, C, and T. In effect, a gene is a long stretch of DNA with its own unique sequence of these four letters. Furthermore, it was found that genes do not exert their influence on the body directly. Instead, chemical intermediates translate the information within each DNA sequence into a molecule of protein.

Proteins perform a variety of roles for the body. Some work as the bricks and mortar of life, giving structural integrity to cells and connective tissues, while others are enzymes—catalysts that speed up chemical reactions within the body.

By the 1960s, genetic knowledge was accumulating at an ever-increasing pace. For developmental biology, a key moment came with the discovery that genes could be turned on and off like light switches. Although every cell in the body carries an identical set of genes, any one cell will use only a subset of the total. Genes that are switched on produce protein, while genes that are turned off remain silent and produce no protein.

It became clear that cells assumed different identities because they employed different batteries of active genes. Any one cell type will make only the proteins appropriate to its function. Skin cells, for example, produce lots of keratin, a protein that gives strength and elasticity where it is most needed—at the body surface. So skin cells have their keratin gene switched on. But genes not relevant to the skin's function—such as hemoglobin genes—remain switched off. Conversely, blood cells produce lots of hemoglobin—the protein that carries oxygen around the body—but no keratin.

The ability of genes to be turned off and on could account for the range of cell identities. But the deeper question still remained: Who was throwing the switches in the first place? Who was overseeing and organizing the whole operation? Who was the architect?

In 1946, Ed Lewis launched himself into an exhaustive study of the fruit fly building site. Lewis was one of the second generation of fruit fly geneticists. A postgraduate student of Alfred Sturtevant, his career at the California Institute of Technology had taken off just as Morgan's was beginning to wind down.

Lewis devoted his attention to *bithorax,* an unusual fruit fly mutant in which the organization of the biological building site had gone awry. Fruit flies, like all members of the fly family, have a single pair of wings. The second pair of wings, found in most other insect groups, has evolved into a pair of gyroscopic balancing devices called halteres, which look like tiny chicken drumsticks. At least, that is the situation in normal flies. But Lewis noticed that *bithorax* had an extra pair of wings where the fly's halteres were supposed to be. On closer inspection, he discovered that it was not just the halteres that had gone missing. The entire body segment to which they were normally attached had been replaced by a duplicate copy of the segment that sat in front of it.

This kind of modification, in which one part of the body is changed into the likeness of something else, is called "homeosis," and it turns up throughout the animal kingdom. There are homeotic crabs, with antennae instead of eyes, and homeotic moths, with legs instead of wings. There are even homeotic humans, although the results are rarely so spectacular. A friend of mine, for example, boasted for many years of his "third nipple." After much goading and several pints of beer, he was eventually persuaded to deliver the evidence one night in a pub. Half-expecting some kind of fleshy breastlet to flop out of his unbuttoned shirt, I was most disappointed when he directed my attention to a molelike affair about fifteen centimeters below his entirely conventional left teat. Spectacular or not, these supernumerary nipples are the result of homeotic mutations.

The *bithorax* fly was unlike most of the mutants that Morgan had studied. Morgan's mutants deviated only slightly from the norm. Eyes changed from red to white, for instance, or bodies changed from brown to yellow. In contrast, the *bithorax* mutation brought about wholesale changes to the body plan of the adult fly.

Despite the difference in the scale of change, Lewis made the startling discovery that the appearance of *bithorax* also boiled down to a modification in a single gene, just like Morgan's simple mutants. There was no doubt that the building of a fruit fly body segment required many different cell types and the action of hundreds of different genes. But Lewis had uncovered a master gene, a molecular architect that could organize and coordinate the building work. The mutant form of the gene behaved like an architect who, having turned up for work drunk, had decided that he was going to install an extra kitchen on the first floor, where a bathroom was supposed to be.

The molecular details of Lewis's discovery turned out to be frighteningly complicated. At least, that was the impression you got when reading his 1978 paper in *Nature,* which summarized over thirty years of work on *bithorax.* The paper was epic in more ways than one, for it broke new ground in its sheer volume and density of jargon. Here's a taste:

A substance, S_o, effecting LMS→LMT, will be assumed to be the Ubx^+ product. The inability of MT to achieve LMT in Ubx hemizygotes or homozygotes is then consistent with the expected reduction in amount of S_o in that segment.

Digesting more than two sentences at a time brought on a feeling of disorientation, dizziness, and nausea in the reader. It was as unreadable, perhaps, as anything yet witnessed on the printed page.

None of this seemed to bother the editors of *Nature*, who, perhaps through some mystical divining or simply through sheer guesswork, decided that this was groundbreaking stuff that deserved to be published. It turned out to be a wise choice. After the jargon had been decoded, what remained were some truly startling biological insights.

Lewis had found evidence for not one but several master control genes arranged in a cluster along one of the fruit fly's chromosomes. The cluster, named the "*bithorax* complex," controlled the development of the rear half of the fruit fly's body. Each gene within the cluster worked like a molecular address within the fly. One gene might say "pair of wings," for example. In cells where this gene was active, a suite of subordinate genes would leap into action, as cells became directed toward the development of a body segment with a pair of wings. Another gene might say "tip of the tail." Cells in which this molecular architect was active would be directed to develop a body segment at the tip of the tail.

A few years after the publication of Lewis's seminal paper, a second gene cluster turned up and filled in the gap left by the *bithorax* complex. The *Antennapedia* complex controlled the development of the front end of the fruit fly's body, and worked in a similar way to its other half.

One of the most bizarre features of these two clusters was that the linear order of genes corresponded exactly with the linear order of body parts. The gene at the head of the queue in the *Antennapedia* complex, for example, controlled the development of the tip of the head, while those behind it controlled body sections progressively farther behind this point. This continued all the way to the gene bringing up the rear of the *bithorax* complex, which controlled the tip of the tail. The significance of this linear correspondence is still unclear because a fly can develop perfectly normally when the order of these genes is disrupted. Perhaps it is an evolutionary hangover from the past, the legacy of a time when gene order was more important to successful development.

As Lewis's paper was going to press, two European biologists, Christiane Nüsslein-Volhard and Eric Wieschaus, were embarking on a mammoth project of their own. Lewis had worked with genes that were active during adult development. But Nüsslein-Volhard and Wieschaus wanted to get back to basics. They wanted to identify genes involved in the early development of the fruit fly embryo, from the fertilized egg to the fully fledged maggot.

Nüsslein-Volhard and Wieschaus were not just interested in finding one or two genes. They wanted to identify all the genes controlling early embryonic development, in a rigorous and systematic way. Their proposal hinted at a special kind of madness that often goes hand in hand with the truly great scientific mind. With no idea of how many genes might be involved, it was impossible to predict just how big their task was going to be.

Undaunted by their critics, Nüsslein-Volhard and Wieschaus plowed on. Their plan was simple. They would feed mutagenic chemicals to thousands of adult flies in the hope of creating a vast repertoire of fruit fly mutants. Individually, any one fly would carry only one or two new mutations. But, taken together, the collection of flies would represent mutations covering the entire set of fruit fly genes. By looking at which of these mutations caused things to go seriously wrong during development, they hoped that it might be possible to deduce what happened when things go right and piece together a sequence of developmental events that led from egg to embryo.

Inside their tiny laboratory in Heidelberg, Nüsslein-Volhard and Wieschaus sat opposite one another at a small table, scrutinizing the production line of mutant fruit fly embryos that passed across the stage of their specially designed dual microscope. For an entire year they remained virtually stationary, as thousands of flies moved across their shared field of vision.

Their dedication to the scientific cause was extreme, although Nüsslein-Volhard, perhaps, did not see it quite that way. Thinking back to those times, she once recalled:

> It was a very difficult but very exciting task. It also was great fun, as so many interesting discoveries were made.

Nüsslein-Volhard was right on one count, at least. Their results were very interesting. Initially, they found fifteen mutant genes that turned the fruit fly building site upside down. Taken together with the homeotic genes, they provided an outline of how genes worked together to organize and control fruit fly development.

To understand the overall picture of genes and development, think of the fly's body in terms of everyday geography. Instead of the body, for example, think of a map of the United States. At the beginning of development, there is just a basic outline of the country. Then a group of control genes swings into action, dividing the outline into north, south, east, and west. A second group of genes, the "state" genes, if you like, is responsible for directing the division of the country into fifty states. Of course, the same "genes" will be present in all states. But in Texas, only the "Texas" genes are switched on, while in

Maine, only the "Maine" genes are switched on. Next the "county" genes become active, dividing each state into a collection of counties. After counties, yet another group of control genes directs the formation of towns and cities within each county, and so on.

Development involved the hierarchical action of different groups of master control genes, which combined in sequence to organize the progressive regionalization of the body. As a fertilized egg, the fly was an amorphous oval, with few distinctive landmarks. As development proceeded, the fly acquired a head end and a tail end, a top and a bottom. Gradually, the body divided up into a series of segments. Only later would the segments assume their distinct identities, under the direction of the homeotic genes.

Lewis, Nüsslein-Volhard, and Wieschaus had added some important pieces to the developmental puzzle, and a coherent story was beginning to emerge. Cells were different from one another because they had different sets of genes switched on, and suites of genes were turned on and off by master control genes.

In 1995, their achievements were recognized formally when all three biologists were jointly awarded the Nobel Prize for medicine and physiology. That it came seventeen years after the publication of Lewis's seminal paper was, perhaps, a reflection of the time it took the Nobel committee to decipher his coded prose.

The discovery of the fruit fly control genes prompted a developmental biology feeding frenzy. The early 1980s saw the emergence of many new and fancy tools in molecular biology, making genetic manipulations quicker and easier than ever before. It was now possible, for instance, to isolate and clone individual genes, and to work out a gene's sequence of DNA letters. Suddenly, everyone was on the hunt for master control genes.

Going hunting for genes might not have the primeval allure of fishing or shooting, but there is still the frisson of excitement that comes with the thrill of the chase and the uncertainty of success. For your average molecular biologist, cooped up indoors for hours on end, hunting for genes is about as exciting as it gets.

Before you can start hunting, you need to know a little bit about your quarry. Genes consist of DNA, a double-stranded molecule. A single strand of DNA is made up of a long line of DNA letters: AATCGGTATTCCA, for instance. Given the sequence of letters on one strand, we can automatically work out the sequence on the other, *complementary,* strand, because the letters form intimate molecular partnerships. The letter A always pairs with T, and G always pairs with C. So if part of one strand has the DNA sequence ATCG, for instance, the complementary strand would have the sequence TAGC.

Although DNA normally exists in its double-stranded form, the two strands can be made to unzip simply by turning up the heat. In other words, DNA works just like the zipper on your trousers. When things get hot, the zipper comes undone; when things cool off, the two strands zip back together again. But two strands of DNA can zip together even if they are not exactly complementary. Provided that most of their DNA letters correspond with one another, they will quite happily zip back up.

This willingness of single-stranded DNA to pair up with a "foreign" strand sharing a similar sequence is crucial when it comes to fishing for genes. To fish, you need bait, and genetic bait comes in the form of single-stranded DNA. The bait is not just any old sequence of DNA. It is made from a gene of the type that you are hunting. The rationale is simple: genes that share similar functions—acting as control switches in development, for instance—usually encode similar kinds of proteins and have similar DNA sequences. So the idea is to use single-stranded DNA from one gene as the bait to try to "catch" other genes that share similar sequences. The bait is lowered into a pool containing a mixture of all of an organism's genes. Under the right conditions of temperature, the bait will zip back up with any genes that share a similar DNA sequence.

Actually obtaining the bait in the first place can be a painstakingly laborious business. To be the first to isolate and sequence a novel gene is always the hardest part of any hunting trip. In the case of fruit flies, it took years to obtain the first DNA sequence of a control gene.

But once identified, the bait proved very effective. Initial fishing expeditions came back with bountiful harvests of genes. What is more, all these genes had the same stretch of DNA sequence in common. It was a DNA "motif" that had been seen before, not in flies, but in bacteria, viruses, and yeasts. And it was characteristic of a type of protein that could dock, *Star Trek* style, with the DNA molecule and turn other genes on or off.

Here was confirmation that the genetic catches were control genes, which could attach to the DNA molecule inside the cell nucleus and orchestrate the action of subordinates. The characteristic DNA motif was named the "homeobox," and it became an instantly recognizable DNA badge that identified a gene as a control gene.

Drunk on success, biologists extended their fishing trips

into foreign waters. They went hunting for control genes in species other than fruit flies, but armed with the same fruit fly bait. And soon they were landing homeobox genes from all over the place. They were found in centipedes, earthworms, fish, frogs, mice, cows, and humans. As the search was extended to embrace more and more disparate elements of the animal kingdom, it soon became clear that the homeobox genes were ubiquitous. Even plants are the proud owners of a neat set of homeobox genes.

One thing has become clear. Developmental genes are an extremely ancient and conservative bunch. Hundreds of millions of years ago, evolution found an effective way of forming a head, a trunk, and a tail. Barring a few minor alterations, it seems to have stuck to it ever since. Despite the dazzling variety of body shapes, it looks as though much of life is built using the same basic rules of design.

So similar are some human and fruit fly developmental genes that the human genes can work just as well in flies as they do in humans. You can knock out a fruit fly homeobox gene and replace it with the human equivalent, and the fly will develop and grow quite normally. Despite being separated by over 500 million years of evolution, it seems that some developmental genes have barely got off their backsides.

Given the similarity of developmental genes, why do we not see human incarnations of the extreme fruit fly mutants? Humans are more complicated and perhaps less robust than flies to wholesale rearrangements of their body plan. But human equivalents do exist. It's just that the outcome is not the comic-book, cut-and-paste image of the muddled man in my dream. Like the fruit fly mutants, many of them die before their development has had a chance to get going. Most mutations in human control genes send the human building site into chaos early on in development. Of the 50 percent or so of human conceptions that end in spontaneous abortion, many are probably due to mutations in control genes.

Not all control-gene mutations have such disastrous effects, but none of them is a bundle of laughs, either. A mutation in the human equivalent of the fruit fly gene *paired* results in an unpleasant condition called Waardenburg's syndrome, the symptoms of which include deafness and defects in the facial skeleton. Another mutation in the human gene *Aniridia* causes a complete loss of the iris.

As their name suggests, and their effects can testify, control genes have a crucial role in development. But they are not the be-all and end-all of the developmental story. Control genes answered the question of how suites of genes are turned on in cells. But in doing so, they created a new problem of their own. If control genes regulate suites of subordinate genes, what is controlling the control genes? What decides when these genes are turned on and off?

Developmental biology was in danger of entering an infinite loop of genetic cause and effect. What was controlling the genes that were controlling the genes that were controlling the genes that were . . . ? But it has been liberated from this never-ending spiral by the return of an old favorite. Chemical gradients are back in the biological mainstream.

Biologists now acknowledge that chemical gradients kick-start the control genes in fruit fly development. Long before any sperm comes into the picture, gradients form at right angles to one another along the three axes of the fruit fly egg. The local concentration of chemical along a gradient is communicated to dividing cells within the egg, and, acting under these instructions, cells begin to form basic positional land-marks (front/back, top/bottom, left/right).

Embryonic development is a complicated business. In some ways, it is like a musical performance by a band or orchestra. Both rely on the interaction of a large number of constituent parts. The finished piece, be it a symphony, a fruit fly, or a human being, will have the form that was originally intended only if each part acts and responds correctly to all the other parts. A broken string or a bum note is unlikely to threaten the whole piece. But if the drummer comes in too soon, or the conductor storms off in a huff, the whole edifice is likely to come tumbling down.

You could argue that comparing the embryonic develop-ment of flies to that of humans is a bit like comparing the Monkees to Mahler; and you might be right. But when it comes to understanding the fundamentals of music—rhythm, melody, and harmony—the three-minute pop song can tell you as much as the glorious polyphony of a Mahler symphony. Don't knock flies. They're the young generation. And they've got something to say.

3

ONE-WAY TICKET

On the western slopes of the Sierra Nevada, somewhere among the oaks, pines, and cedars of a pristine California forest, a translucent tube, no bigger than a mouse dropping, is about to reveal its contents. With its development complete, a wild fruit fly is ready to leave its pupal case and embark on a short but hectic adult life.

The heat of the late afternoon sun sends a fault line careering through the crispy pupal husk, and a moist, shiny body climbs up and out, toward the sky. Jerky prototype movements take the fly to a suitable perch on the tree, where it hangs itself out to dry.

After a period of quiet and motionless contemplation, the fly emerges from its Zen-like state with a violent twitch. As it roams over the tree bark, its movements are now assured and precise. Orientating itself to the more demanding dimensions of the adult world, the fly leaps into the air and heads off on its maiden flight.

The fly pauses at the fringe of the forest, where stands of adolescent trees rub shoulders with a patch of open meadow.

With the sun dipping below the tree line, the meadow's mix of color is beginning to ebb away, the pastel shades dissolving into a uniform expanse of muddy brown. Sunset seems like an ideal time to set off in search of carnal and culinary pleasures. As shadows spill over into night, the young fly quietly slips away into the darkness.

The next morning, Russian-born biologist Theodosius Dobzhansky wakes early. He eats a quick breakfast of eggs and black coffee and then heads off to inspect his traps. He's in a good mood. It's great to be away from the city again, spending some time outdoors.

In recent years, a small, isolated log cabin on the fringes of Yosemite National Park had become a second home to Dobzhansky. Out in the wilderness, he could indulge his twin passions of camping and fieldwork. For him, this was as close to heaven as you could get, without actually paying the admission price.

In the meadow, Dobzhansky walks slowly and deliberately from trap to trap. He carefully inspects the contents of each one before moving on to the next. It must have been a quiet night. All the traps are empty. All, that is, except one. Dobzhansky wanders over to where the meadow merges into woodland. Peering inside the trap, he smiles at the sight of a young male fly, writhing around in a glutinous paste of rotting banana.

Back in the 1930s, Dobzhansky was a fruit fly radical. He abandoned the domesticated *Drosophila melanogaster*, star of the Fly Room and a million laboratory crosses, for *Drosophila pseudoobscura*, a wilder and less well-known fruit fly cousin. In doing so, he helped to build a bridge between two opposing biological traditions.

It has always been convenient, if sometimes simplistic, to divide biologists into two distinct camps, namely the experimentalists, as exemplified by Morgan, and the natural historians, as exemplified by Darwin. The legacy of this divide can still be seen today. Modern biologists, like belly buttons, tend to fall into one of two categories, "innies" or "outies." "Innies," the modern descendants of the experimental tradition, spend their entire working lives indoors. They are most comfortable sitting at the computer or laboratory bench and develop acute migraines when exposed to direct sunlight. To this group belong the biochemists, molecular biologists, geneticists, and mathematical modelers. Most of these people will not own a pair of binoculars.

In contrast, "outies," the modern-day naturalists, are laboratory-illiterate. They understand how to open the fridge door, but that's about as far as their indoor knowledge extends. None of this matters to an outie, of course. Outies are more interested in devoting all their energy to prodigious beard growth and memorizing the Latin names of a thousand different bird species. All outies own a pair of extremely expensive binoculars, which are worn at all times, with the maker's name facing outward. To this category belong the ecologists and, well, that's about it.

Occasionally, however, you come across a third category of biologist, someone who is neither an innie nor an outie, but an "in-betweenie." These are the rare individuals who feel equally at home in both artificial and natural light, people who can distinguish Petri dishes from pelicans, and can somehow manage to incorporate both into a single experiment.

Dobzhansky was perhaps the first and certainly the greatest example of an in-betweenie. He divided his time equally between the laboratory and the field. In the summer he would collect and study fruit flies in the wild, bringing them back to the laboratory for further scrutiny in the winter. His break with tradition was a deliberate move designed to take biological experiments beyond the confines of the laboratory and into the great outdoors.

Dobzhansky's new mode of scientific practice was not only revolutionary, it was also extremely productive. By moving outdoors with the fruit fly, he became instrumental in the modernization of evolutionary biology. Combining elements of both the experimental and naturalist traditions, he united genetics with Darwinian evolution to help forge the new science of evolutionary genetics.

It all could have been so different had Dobzhansky decided to stay in his native homeland rather than move to America. Had he done so, he would almost certainly have suffered the same fate that befell scientific contemporaries such as Nikolai Vavilov, and the millions of other Soviet citizens to whom Stalin took a personal dislike.

As it was, he arrived in the United States in 1927, aged twenty-seven, to join Morgan's group in New York. His original intention was to stay for a year—the duration of his Rockefeller fellowship—then return home to Russia to set up a fruit fly genetics laboratory at Leningrad University. But with the political climate in the Soviet Union worsening, his stay became permanent.

Politics notwithstanding, the move to Morgan's laboratory was a dream ticket for Dobzhansky. He had read all about Morgan's work in the early 1920s. "That was a sort of a revelation," he once remarked, "... by that time, genetics was *it*. Morgan was a hero or a saint." Hero or not, when he arrived at Columbia University, Dobzhansky was appalled by Morgan's workspace. The dirt and general disarray of the Fly Room, the cockroaches in the desk drawers, the nose-worrying smells, the constant noise of clanking bottles; it was not what he had expected from a scientific idol.

Though Dobzhansky was a great fan of Morgan's work on genetics, the two men had completely different scientific outlooks, particularly when it came to evolutionary biology. Morgan tended to look down his nose at evolutionary ideas. As far as he was concerned, evolutionary biology, being an archetypal product of the naturalist tradition, was largely speculative and nonscientific.

Morgan's view was typical of a self-proclaimed experimental biologist brought up in a culture that divided biology down the middle. But it was a view completely alien to Dobzhansky. The experimentalist/naturalist dichotomy was very much a Western invention. No such entrenched division existed in the East. Dobzhansky, who had been trained in both experimental science and field biology, saw no conflict between the two, and he found Morgan's hostility and snobbery toward evolutionary and field biology difficult to stomach.

Considering his frequent objections to evolutionary ideas, it is intriguing that Morgan chose to write three books that had "evolution" in the title. All the books reveal the recurrent hang-ups that Morgan felt about natural selection. In his final book on the subject, *The Scientific Basis of Evolution,* published in 1932, Morgan seemed convinced that natural selection was dead and buried. At one point in the book, he unleashes the bold assertion that "the implication in the theory of natural selection that by selecting the more extreme individuals of the population, the next generation will be moved further in the same direction, is now known to be wrong."

Quite where Morgan was getting his information is unclear, but his sources were woefully inaccurate. By 1932, Darwinian natural selection was undergoing a renaissance, and Morgan's was definitely the minority view. Evidence, accumulating from a variety of sources, was showing that selection could lead to directional changes in the visible features of plants and animals. In the laboratory, all sorts of characteristics, from the width of a pigmented stripe on the back of a hooded rat, to the number of hairs on a fruit fly's body, were being shown to be amenable to selection. Out in the "real world," too, the experience of successful animal and plant breeders was also lending credence to the malleable powers of selection.

Morgan's inability to see eye to eye with natural selection stemmed from his fundamental misconception of genetic variability in natural populations. Although he acknowledged that the kinds of mutant flies that turned up in his laboratory could also appear in nature, he considered these flies aberrations from an (idealized) "wild type." Natural populations, he believed, were effectively homogeneous collections of genetically identical individuals. By denying the existence of heritable differences between individuals, it is no surprise that he also denied natural selection a meaningful role in evolution.

Maybe Morgan had spent too much time cooped up indoors. Had he got out more, he might have been more willing to accept what the naturalists had been saying for years. Namely, that populations are full of variability. Take one hundred fruit flies from the wild and measure whatever feature you like—eye color, head width, penis length, or the number of hairs on their backside—and for each of these characters, and hundreds more, there will be measurable differences between individuals. Measure enough characters and it will soon become obvious that every individual in the population is unique.

In his defense, Morgan's view was not quite as myopic as it sounds. Before 1933, there was no way—short of suicidally intensive crossing experiments—of measuring *genetic* variability at the population level. Sure, there was plenty of variability in populations. But this was not necessarily the same thing as genetic variability. After all, genes are not the only reason why individuals differ from one another. The environment can also make a contribution. Think of Americans, for example, who are, on average, heavier than British people. This difference is not due to genes. Americans just eat more.

In complete contrast to Morgan, Dobzhansky eagerly devoured all things Darwinian. In the Soviet Union, Darwin's ideas had become far more embedded in the popular culture than they had in the West. Dobzhansky claimed to have read *On the Origin of Species* when he was thirteen. By his early twenties he was spending his summers studying variability in wild populations of Russian ladybirds. The move to Morgan's laboratory, to learn about the new genetics, would, he hoped, provide him with a more refined evolutionary understanding.

But mathematicians were already one step ahead of him. Since the early 1900s, they had been muscling in on evolutionary ideas, shifting genetics from the study of individuals to the study of genes in populations. In the 1920s, Sewall Wright in the United States, and Ronald Fisher and J.B.S. Haldane in the United Kingdom, were incorporating the simple mathematical rules of Mendelian genetics into theoretical studies of populations.

For most biologists, the implications of these mathematical models lay outside their understanding. This was no great surprise. Biologists have always had a troubled and confused relationship with mathematics. For your average biologist, the mere sight of an algebraic expression can be traumatic. Many people who want a career in science choose biology just to avoid the mathematics that plagues physics and chemistry.

But even in biology, it is difficult to dodge mathematics entirely. Thanks to people such as Fisher and Wright, evolutionary biology is riddled with the stuff. I can still remember the bewildering experience of "reading" some of their seminal evolutionary papers. On confronting the first equation, I would study it earnestly, as if stumbling on some new and mysterious archaeological find. I would then try to convince myself that if only I concentrated hard enough, the true meaning of this abstract gobbledygook would appear in a revelatory vision. Of course, the vision never came. So I would move on. As more and more text turned to algebra, my exasperation would grow until, at some peak point of frustration, I would fling the paper to one side, onto an ever-growing pile marked "to be read later."

Even scientific "greats" such as Morgan fought shy of evolutionary mathematics. Dobzhansky, too, had trouble coming to grips with it. But, unlike Morgan, he did persevere. Dobzhansky once remarked:

> I am not a mathematician at all. My way of reading Sewall Wright's papers, which I still think is perfectly defensible, is to examine the biological assumptions which the man is making, and to read the conclusions which he arrives at, and hope to goodness that what comes in between is correct.

Mathematicians were creating an entirely new way of looking at evolution. Instead of thinking about populations of plants and animals as collections of individuals, they were thinking exclusively in terms of genes and "gene pools." Populations were modeled as bags of genes, whose frequencies changed according to their selective advantage. Any gene that provided some kind of competitive edge in the struggle for existence would become more common in a population.

Dobzhansky was only too aware of how important these mathematical studies were. They signaled the emergence of a theoretical framework, thus far absent from evolutionary biology. All that was needed was someone who was prepared to do what few Western experimentalists had dared to do: to break free from their laboratories, get out into the countryside, and put the theory to the test.

In 1928, Morgan moved his entire research group to the California Institute of Technology in Pasadena, on the outskirts of Los Angeles. The change suited Dobzhansky, whose love of traveling and camping was perfectly accommodated by the vast open spaces of the California countryside. It was through his summer excursions into the California wilderness that Dobzhansky became acquainted with *Drosophila pseudoobscura,* the fruit fly that would later shape his career.

For Dobzhansky, the switch from one *Drosophila* species to another was born out of necessity. By the early 1930s, the laboratory stalwart *Drosophila melanogaster* had become so domesticated that it was virtually toilet-trained. Any semblance of the natural had disappeared long ago, when the species abandoned the wilderness for the more comfortable human surroundings of dustbins, wine cellars, and fruit orchards. Also, populations living outside the laboratory could have been "contaminated" by individuals that had escaped from the inside. Dobzhansky decided that he would need to look elsewhere if he wanted to study the genetics of truly natural populations.

Drosophila pseudoobscura was an ideal replacement. It was wild but accessible: populations lived within easy reach of Pasadena. (Its geographical range covers the western half of North America and extends into Mexico. There is also a single isolated population near Bogotá in Colombia that, presumably, is the subject of close scrutiny by the Drug Enforcement Agency.) Like its more established laboratory cousin, it was unfussy in its habits and could be made to feel quite at home in the laboratory.

But *Drosophila pseudoobscura* had other, less obvious, features that made it particularly suited to Dobzhansky's needs. Genetic variation in this species was not just about different versions of individual genes. It was also about different orders of genes along a chromosome. One version of the chromosome might have the gene order ABCDEFG, for example; another might have the order ABEDCFG. Still another might have the order ABDCEFG, and so on. These chromosomal variants were the product of mutations, called inversions, in which a fragment of chromosome, containing a string of genes, is turned back to front.

Inversions were identified in the early days of the Fly Room. Originally, they were difficult to detect because their presence could be inferred only through painfully laborious crossing experiments. But in 1933, the situation changed completely when fruit flies served up a surprise in their salivary glands.

The chromosomes inside the fruit fly's salivary gland cells are huge—a thousand times thicker than normal. No one knew it at the time, but the thickness is due to the chromosome's DNA replicating many times over, without the cell dividing, so that each chromosome is like a packet of spaghetti. Chemical staining of these supersized chromosomes reveals dark bands along their length, distinct landmarks corresponding to the positions of specific genes.

The salivary-gland chromosomes were a godsend to Dobzhansky. He could look at the banding patterns under a microscope and easily distinguish the different chromosome inversions. These chromosomes, with their distinctive banding patterns, became biological bar codes that were used to provide the first reliable measures of genetic variation within populations.

By the mid-1930s, Dobzhansky was traveling extensively, collecting flies from throughout their range. He went south into Mexico, north into British Columbia and Alaska, and east as far as Nebraska and North Dakota. Thousands of flies were brought back to Pasadena to have their chromosomes scrutinized under the microscope.

With Dobzhansky churning out results in earnest, it became immediately obvious that the old (and by now extremely rusty) idea of populations as collections of genetically identical individuals was destined for the dustbin of history. Every population that Dobzhansky looked at contained a reservoir of genetic diversity. As he, and many others, had suspected, genetic variation was no aberration; it was a fact of life.

To illustrate the kinds of patterns that Dobzhansky observed, it might be easier to think in terms of something other than a chromosome. Imagine, for example, that Dobzhansky was studying shoes on the feet of people in different U.S. cities. The basic principle is the same. A shoe, like a chromosome or a gene, not only comes in a pair; it also comes in a variety of different types, and types vary between individuals. In the interests of fashion, the two shoes in a pair are usually the same. But for the purposes of this analogy, we will assume that they can also be different. The key is to think in terms of the frequency of different shoe types in populations, and forget about pairs of shoes on individuals.

As I mentioned above, the first thing Dobzhansky found was that there was lots of genetic variation within each population. In shoe-speak, there were about half a dozen different types of shoes in each city. But not only was there variability within populations, there was also variability between populations. In other words, different cities would have different shoe-type profiles. Roman sandals and moccasins, for example, were popular in San Francisco but rare in Minneapolis, where snow boots were all the rage. In Dallas, both Roman sandals and snow boots were difficult to find, but cowboy boots were everywhere. Designer sneakers were popular in Los Angeles but rare in Seattle, where rubber boots predominated. In contrast, New York displayed a roughly even mix of all shoe types, including leather oxfords, which were difficult to find elsewhere.

The difference between two populations of flies depended on the geographical distance between them. In the language of footwear, the shoe profiles of Los Angeles and San Francisco were more similar to one another than those of Los Angeles and New York. But against the background of these broad geographic patterns, Dobzhansky also noticed some fine-scale detail. Just as shoe profiles in Venice Beach would differ from those in downtown L.A., he found that nearby fly populations could still be genetically distinct if they occupied different habitats, such as a meadow and a forest, or different altitudes on a mountain slope.

In extreme cases, genetic differences could lead to reproductive incompatibilities. When, for instance, a female from a Colombian population of *Drosophila pseudoobscura* was mated to a male from a population in North America, the male offspring were sterile. In this instance, reproductive isolation was not complete—female offspring were fertile. But this kind of evidence convinced Dobzhansky that small, incremental genetic changes could eventually give rise to reproductive barriers. These reproductive incompatibilities, Dobzhansky believed, defined the boundaries between species.

In accumulating genetic differences, he saw how two populations might also accumulate differences in body size, color, genital architecture, behavioral idiosyncrasies, and a thousand other characteristics that could eventually make them reluctant or unable to mate with one another. In these distinct genetic profiles, Dobzhansky believed he was seeing the origin of species in its infancy.

Many of the profiles that Dobzhansky studied were not static, but could change over remarkably short timescales. When he began taking repeated monthly samples of flies from a population on Mount San Jacinto in California, for example, he found that the frequency of some chromosome types went through yearly cycles. In the context of shoes, snow boots might reach peak popularity in the Minneapolis winter, and then decline in the spring and summer, when moccasins reach their peak. As autumn approaches, moccasins decline, as snow boots increase once again.

What was the cause of these seasonal fluctuations? Initially, Dobzhansky attributed the changes to chance, to the random element in inheritance. But he returned to the same population each year and recorded the same pattern every time. When he realized just how regular the changes were, random forces had to be discounted. There was really only one alternative explanation for these fluctuations, and that was natural selection. In the struggle for existence, it seemed that specific chromosome types, like specific types of shoes, did better at certain times of the year than at others.

It is not going over the top to say that the results from Mount San Jacinto were epoch-making for evolutionary biology. Because evolution had traditionally been considered a slow-paced affair that was difficult, if not impossible, to test experimentally, critics had dismissed the subject as unscientific. But here, on Mount San Jacinto, was a perfect demonstration of evolution in action. This was no million-year wait for some pathetic two-millimeter increase in the length of a leg bone. This was evolutionary change in front of your very eyes.

Dobzhansky's studies gave biologists renewed faith in the power of natural selection. Through his work with the fly, he had successfully brought to life what the theoreticians had, more or less, predicted. Provided that natural selection was strong enough, it seemed that anything was possible.

In 1937, Dobzhansky seriously damaged his knee in a horse-riding accident. Never one to laze around, he used his recuperation time to write a book, with the explicit intention of bringing evolutionary biology up to date. The result, *Genetics and the Origin of Species,* was an instant classic. In effect, Dobzhansky used the book to present a synthesis of the theoretical and the empirical, molding evolutionary mathematics and the latest experimental observations into one cohesive whole. In doing so, he assumed the role of interpreter, somehow managing to make the abstruse mathematical theory more palatable and digestible to frightened biological minds. He also guaranteed that chromosome inversions of fruit flies would keep evolutionary biologists happy for years to come.

Drosophila pseudoobscura may have been helping to unify genetics and evolutionary biology out in the field. But back in the laboratory, it was propping up a rift of Grand Canyon–esque proportions between Dobzhansky and a fellow member of the fly group, Alfred Sturtevant.

At one time, Dobzhansky and Sturtevant had been close colleagues and friends. When Dobzhansky first arrived in the United States, it was Sturtevant who took him under his wing and showed him the ropes. It was Sturtevant, too, who had introduced Dobzhansky to the pleasures of *Drosophila pseudoobscura*. They had shared a laboratory and collaborated closely on a number of projects with the fly.

Both of them were interested in the application of genetics to evolution, but they approached the issue from completely different perspectives. Sturtevant wanted to be the first to use genetics to decipher evolutionary relationships between different fruit fly species. Dobzhansky, on the other hand, was more interested in the mechanism of evolution than its products. He wanted to focus on the origins of genetic differences between populations of the same species.

On the surface, at least, their differing ambitions presented little evidence of conflict. Early in 1936, both men were talking about a big collaborative project on the genetics of fruit flies, a project that would unite the interests of both. But trouble was bubbling beneath the surface. And in May of that year, it finally erupted, sending their friendship into free fall.

It all started when Dobzhansky was offered a professorship at the University of Texas. The offer was a prestigious one and reflected the high regard in which he was now held by the academic community. When Morgan heard the news, he immediately made Dobzhansky a counteroffer. Unsure of what to do, Dobzhansky went to his friend Sturtevant for advice.

Sturtevant made it clear that he would be mad not to accept the Texas job. It offered the prospect of generous quantities of cash, laboratory space, and research staff. It was also a great opportunity for academic independence. So Dobzhansky accepted the job. But after mulling it over for a while, he changed his mind, and wrote to the University of Texas, telling them he would be staying in Pasadena.

When Sturtevant heard that Dobzhansky was staying, he found it impossible to conceal his disappointment. "His face fell," Dobzhansky recalled. "It was quite obvious he didn't want this. That was a terrific shock." Stunned by Sturtevant's reaction, Dobzhansky immediately wrote back to the University of Texas, saying that he had changed his mind again and now wanted the job. But the post was no longer available.

From his reaction, Sturtevant was obviously hoping to see the back of Dobzhansky, although the reasons were never fully resolved. Perhaps he was sick of Dobzhansky's rough-and-ready work style, and his obsession with productivity. Dobzhansky was fond of saying that a month gone by without a paper sent to press was a wasted month. That kind of statement would be irritating enough for your average academic, let alone someone like Sturtevant, whose approach to science was always meticulous and methodical.

Maybe his reaction had more to do with their differing scientific interests. Sturtevant obviously had great faith in his own brand of evolutionary genetics, and saw it as a way of establishing a more concrete scientific identity for himself. Perhaps he just didn't want Dobzhansky getting in the way.

But there were plenty of other issues at stake that could account for Sturtevant's desire to see Dobzhansky go. Before the incident, tension in the fly group had been building over Morgan's impending retirement and his reluctance to name a successor. Sturtevant probably thought the job was rightfully his, a just reward for a career sacrificed to the interests of the group. But there were no guarantees with Dobzhansky also in the picture. Dobzhansky was certainly a friend. But he had also become a rival, and, almost ten years Sturtevant's junior, a much younger one at that.

Whatever the reasons behind Sturtevant's reaction, the incident left a festering wound in the fly group and signaled the start of its slow and lingering death. Sturtevant moved out of the laboratory he had shared with Dobzhansky and communication was reduced to the basic civilities. The collective spirit that had characterized the group for so long evaporated as Dobzhansky withdrew to its fringes.

In 1938, the bad atmosphere was not helped by the sudden and tragic death of Calvin Bridges, from a heart attack, at the age of forty-nine. Bridges, like Sturtevant, had been a founding member of the fly group at Columbia University, but there the similarities ended. The personalities of the two men could not have been more different. Sturtevant worked hard to cultivate the image of the intellectual. He was prone to arrogance, and frequently intolerant of those he considered beneath him. In contrast, Bridges saw himself more as a technical expert, a skilled laborer, and people found him extremely likable. He exuded a potent mix of flamboyance, generosity, and gullibility. Dobzhansky once said that he possessed a divine spark.

Bridges was also unconventional. Having studied the fruit fly lifestyle, he seemed to adopt it as his own. In the early 1920s, he left his wife and children, had a vasectomy, and embarked on a crusade of sexual promiscuity. Formal notions of seduction were abandoned in favor of a more direct approach. Perhaps his death was a consequence of his lifestyle finally catching up with him. Or perhaps he just had a weak heart.

With Bridges gone, Dobzhansky remained in Pasadena for another two years. Relations with Sturtevant, meanwhile, continued to go downhill. In 1939, Dobzhansky sounded a depressive tone in a letter to his friend Milislav Demerec: "It is better simply not to care for a person than after many years to find out that he is not worth one's care."

In 1940, he was offered a post at Columbia University, and accepted immediately. For Dobzhansky, the move could not have come too soon. He wrote to Demerec: "... I am dead tired of [the] Pasadena environment, except for the natural environment—mountains, deserts and valleys, which I really love, and the loss of which I shall regret." With Dobzhansky finally on his way, Sturtevant wrote him a letter full of apologies and kind words: "The place will seem strange without you, for none of the rest of us have your energy and go. Yes, you may be sure we'll miss you."

Despite the move back to New York, Dobzhansky maintained his annual migration to fruit fly stomping grounds in the California wilderness. Long, hot summers were spent in the San Jacinto Mountains of southern California and on the western slopes of Yosemite National Park.

Even the Second World War could not distract him from his dedication to the fly. While his compatriots were fighting the forces of fascism, Dobzhansky was preoccupied with the no less important issue of measuring how far flies fly.

Implausible as it may sound, knowledge of an animal's travel arrangements is integral to an understanding of their evolution. When animals move around they take their genes with them. If individuals born in one population move off and reproduce in another, then they act as hereditary whisks, moving genes among populations, and smoothing out the differences between them. On the other hand, if there is little or no movement of individuals, then there is more potential for populations to evolve along separate paths. "Gene flow" is the term geneticists use to describe the movement of genes between populations, but you can just as easily picture it as "shoe flow." The basic idea is simple. The more that people move between cities—the more "shoe flow"—the more similar the shoe profiles of the two cities will become.

Gene flow, I have to confess, is a subject particularly close to my heart. For better or for worse, I spent four years of my academic life studying gene flow, not in flies, but in a small species of moth. The species hardly matters. The principle was exactly the same as that established by Dobzhansky almost half a century earlier.

My field site, a vast coastal sand-dune system in South Wales, did not have quite the kudos of Dobzhansky's California hangouts. But if you ignored the sights and smells of the nearby chemical refineries, it was still a beautiful place to work. Beautiful, that is, during the day. By nightfall, however, the whole character of the place was transformed, as the gentle folds and rolling undulations of the dunes assumed a menacing character. The sense of danger came not from the indigenous wildlife but from the neighboring human population. I was warned by the local police that a few unsavory types were known to migrate onto the dunes, under the cover of darkness, to indulge in what the police described as "various criminal activities." I did not warm to this news. The moths I was studying were most active in the middle of the night. At about 3 A.M. to be exact.

To monitor how far they flew, I began by marking hundreds of moths with fluorescent dust, so that they glowed like fireflies under ultraviolet light. The idea was to release the moths into the wild at specific spots among the dunes and then return on subsequent nights to hunt them down with my ultraviolet lamp.

For the next two weeks, I found myself stumbling around the sand dunes, scouring the darkness for specks of light. I was so concerned about getting burned by the ultraviolet radiation coming out of the lamp that I covered my face with a huge plastic visor, and wore thick gloves and a scarf. From a distance, I looked like an alien from a cheap 1950s sci-fi movie.

But nocturnal sunburn was the least of my worries. I was far more concerned about whom I might meet out there in the darkness. The dunes at night were a quiet and lonely place, especially when the moon was hidden by clouds. Sometimes, strange sounds would interrupt the silence. They came from unidentifiable directions and were like nothing I had heard before or since. Weird electric shrieks and squawks, and long persistent hissings. Were these just the natural sounds of nocturnal living, of territorial disputes and arguments over sex and food? Or were they the sound of "various criminal activities"? On more than one occasion, I sprinted back to the haven of my car rather than hang around to find out.

Perhaps fruit flies, rather than moths, would have been a better option. After all, their habits are not nearly so antisocial. For *Drosophila pseudoobscura,* rush hours are around dawn and dusk. It was at these times each day that Dobzhansky would inspect his array of banana-baited traps to see if any flies had been enticed inside.

Dobzhansky's marking technique was rather more sophisticated than my own. Flies were marked not by fluorescent dust but by a mutant gene. He reared and released thousands of flies with vivid orange eyes, which could easily be distinguished from the wild, red-eyed flies that also turned up in the traps.

Dobzhansky's flies were also more adventurous travelers than my moths. Flies could move up to a hundred meters a day. By contrast, my pathetic little moths were lucky if they moved a few meters in a lifetime. This meant that, potentially, evolutionary change could take place over very small distances. From an evolutionary point of view, moths separated by twenty meters or so belonged to quite separate populations.

Dobzhansky might not have had to get up at 3 every morning, but he, too, suffered from the fear factor. For Dobzhansky, the threat came not in the form of midnight ramblers, but in the far more sinister guise of the Federal Bureau of Investigation. A man with an East European accent wandering around remote parts of California during the war was enough to arouse their curiosity and suspicion.

Who knows what was on the paranoid minds of the FBI agents? Perhaps they believed that Dobzhansky was training flies for reconnaissance missions, to penetrate the heart of the Pentagon and expose state secrets to the enemy. Whatever the source of their suspicion, Dobzhansky underwent a grueling and extremely distressing interrogation.

Unfortunately, I have been unable to get hold of the transcripts, but one can hazard a guess at the general tone and substance of their conversation:

FBI: *How far did the flies fly?*

Dob: About one hundred meters a day.

FBI: *Hmmm . . . The distance from the security gate to the Pentagon's inner sanctum? Not quite. But near enough.*

Have you ever visited Washington, either as a tourist or on business?

Do you enjoy vodka?

Do you prefer the Urals or the Rockies? Tennessee Williams or Anton Chekhov? Where are you getting your supply of bananas for the traps? What is the ultimate purpose of your work?

Dob: To understand the genetics of natural populations.

FBI: *Sounds dodgy.*

Dobzhansky tried to explain the science behind what he was doing. But in front of a skeptical FBI investigator, he had trouble making a credible case for evolutionary genetics. Here I can empathize with Dobzhansky. There were many times when I tried to impress on doubting relatives the importance of my own research. But my attempts were invariably met with contempt and disdain. As my uncle once said, "The money that us taxpayers are spending on those bloody moths could put an extra lane on the M25 between Watford and Staines."

If, like my uncle, you are the kind of person who likes your science to have real practical "benefits" to the human race, then you might not find evolutionary genetics to your liking. No evolutionary geneticist has ever won a Nobel Prize, and measuring how far flies move is unlikely to add much to a nation's Gross Domestic Product. But it would be premature to write it off just yet. After all, the genetics of populations, like the genetics of individuals, is a universal language. The rules that apply to a population of flies are the same as those that apply to populations of moths, aardvarks, and humans—even to a population of cancer cells in a growing, evolving tumor.

But if you want tangible benefits, you can always find them if you look hard enough. Forensic scientists, for example, use population genetics to assess whether a DNA "fingerprint" match between a suspect and the scene of a crime could be due to chance alone. Anthropologists also use population genetics to trace human history. Like languages and cultures, the worldwide distribution of genes can be used to infer the route of human global colonization.

Before biologists like Dobzhansky appeared on the scene, populations were seen as collections of genetically identical individuals. With a little help from his fruit fly friends, Dobzhansky liberated populations from this philosophical straitjacket. In the new, enlightened world, populations were reservoirs of genetic diversity and genes were the currency of evolutionary change. The genetic profiles of populations were kept in a dynamic flux by a mix of interacting evolutionary forces. Darwinian natural selection could mold the genetic profiles of populations by favoring certain types of genes over others. Mutation—the original source of all new variety—could inject genetic novelties into populations. And gene flow could mix the genes of different populations and iron out differences between them. In this new world order, evolution depended on the balance between these interacting forces.

To illustrate the way in which these evolutionary forces can work together, we'll return to the shoe analogy for one last time. Forget historical accuracy and imagine that the population of Minneapolis originated when a group of Dallas natives grew tired of oil and cowboys and decided to head north for pastures new.

If the emigrating population is sufficiently large, then it is likely to carry a representative sample of all the Dallas shoe types. If the splinter group is small, however, then there is a possibility that, by chance, it will carry a biased sample of the Dallas shoe profile. Shoe types that were rare in Dallas are likely to be absent. Other shoe types may be disproportionately common. These kinds of random changes in the frequency of genes, or shoes, caused by sampling bias, are called founder effects, and they are another way in which populations can evolve.

Let's assume, for the sake of this analogy, that the splinter group carries a representative sample of the Dallas shoe types. So, initially, the two populations—Dallas and the Minneapolis founders—have identical shoe profiles. But fairly soon it becomes clear that shoes that were ideal in Dallas are not sensible footwear in Minneapolis. Because of the cooler conditions in the north, "natural selection" favors different kinds of shoes, and the two shoe profiles began to diverge from one another.

If natural selection is sufficiently strong, these differences will be maintained despite shoe flow between the two cities. But if selection is weak, migration may bring the profiles back together again. In this dynamic flux, some shoe types may disappear altogether. Cowboy boots, for example, may not last long in Minneapolis. In these circumstances, only continued migration from Dallas, or mutation, could breathe new life into the Minneapolis cowboy boot. Mutation could also provide an entirely new variety, platform shoes in the 1970s being an excellent example.

Shoe profiles of the two populations may continue to diverge. But even if the inhabitants of Dallas and Minneapolis no longer consider themselves compatible, different tastes in shoes are unlikely to lead to the origin of new species. Shoes, it seems, can take you only so far. Genes can take you that little bit further.

Genetics made evolution and the origin of species more credible to a once skeptical scientific community. It also made a name for a new and lesser-known species of fruit fly, *Drosophila pseudoobscura*. Through a mixture of good judgment and good luck, Dobzhansky stumbled across the kind of experimental system that most biologists only ever dream about.

Darwin, no doubt, was turning in his grave. Throughout his life, the credibility of his evolutionary ideas had been hampered by the lack of any genetic foundation. Genetics not only tightened up his theory, it also turned evolutionary biology into an experimental science. Darwin's evolutionary evidence had come from comparative studies of organisms and their environments, from the fossil record, and from plant and animal domestication. As exceptional as the evidence was, it remained descriptive and indirect. Studies of fruit fly chromosomes added a welcome dose of experimental legitimacy.

Darwin, I'm sure, would have given anything for a share of Dobzhansky's spoils. He might have shaved off his beard and danced naked at a Linnaean Society annual general meeting; he might have declared "I am insane" to a congregation of ardent creationists; he might even have forgone an all-expenses-paid trip for two to the Galápagos Islands, with top-of-the-line beak-measuring equipment thrown in. He might have sacrificed all of this and more had it guaranteed him a share of Dobzhansky's experimental riches. But it wasn't to be.

Serves him right for looking at finches rather than flies.

4

THE SCHOOL OF HARD KNOCKS

The 1970s were a boom time for the fly. For the previous thirty years, it had been forced to play second fiddle to the bacteria and virus brigade. But in the decade of disco, the fruit fly was suddenly hip again. It was as if the flapping of scientific flares had stirred up the laboratory atmosphere, creating bottle-sized whirlwinds that roused the flies from their slumber. Even Hollywood, for once, seemed to be in tune with the Zeitgeist, releasing the big-budget blockbuster *Superfly*, complete with a cool wah-wah soundtrack from legendary soul man Curtis Mayfield.

The epicenter of this scientific renaissance was in Germany, at a small molecular-biology laboratory in Heidelberg. It was there that developmental biologists Christiane Nüsslein-Volhard and Eric Weischaus relaunched the fly's career. But the fly was also making a name for itself elsewhere. Ten thousand miles away, at the California Institute of Technology, the fly was emerging as a key player in genetic studies of behavior.

The architect behind these behavioral studies was the scientific polymath Seymour Benzer. In the early 1940s, Benzer began his academic career as a physicist. Switching to biology in the 1950s, he joined the first generation of molecular biologists and made a name for himself dissecting the molecular structure of the gene. By the 1970s, he had changed direction again, this time turning his attention to the genetic dissection of behavior.

One of Benzer's most important achievements during the 1970s was to bring about a sea change in the popular image of the fly. While it is true that the fly is a highly sexed creature, Benzer put paid to the idea of the fly as a brainless pinhead, a sex-crazed automaton. On the contrary, Benzer's group showed that flies had an intellectual streak. Given the right training, they could learn and memorize information.

Domestic dogs around the world must have been trembling in their baskets as this news filtered out of Benzer's laboratory. With their reputation for intelligence, they had dominated the domestic-pet market for years. But now there was a new kid on the block. If intelligence was measured by the speed with which an animal can memorize information, then dogs made fruit flies look like mini-Einsteins.

Training a dog can take days, weeks, or even months in the case of the more brainless breeds. In contrast, you can train a fruit fly in two minutes flat. Admittedly, a fly does not respond to commands like "sit" and "stay"; but this is hardly surprising. Sitting does not come naturally to a six-legged animal like a fruit fly.

Forget verbal commands altogether. The best way to train fruit flies is with odors and electricity. Here's how you do it. Stick some flies inside a test tube. Pipe a strong odor into the tube while simultaneously zapping the flies with an electric shock. The shock should be about seventy volts—a strong jolt to the system, but not enough to kill them. After a minute, turn off the

power and pipe a second odor into the tube—without a shock—
for a minute more. And that's it: training exercise over.

Next comes the exam, the test to see whether the flies have
learned by association. Remove the flies from their tube and
place them at the choice point of a T-shaped maze. The "test"
for the flies is to decide whether to walk toward odor one—the
odor originally presented with the electric shock—or odor
two—the one without the shock. In an average class of flies,
over 90 percent will pass the test, by walking toward the second
odor.

With only one training exercise, the fruit fly memory is
fairly short term. Repeat the test three hours later and you will
find that some of the flies are already showing signs of forget-
fulness. After twenty-four hours, all the flies will have completely forgotten what they learned in training.

This doesn't mean that flies are incapable of long-term
memory. It's just that flies require repeated training for memories to stick. If you repeat the fruit fly electroshock exercise ten
times, with fifteen-minute gaps between each training session,
then flies will still be able to distinguish the "good" and "bad"
odors seven days later.

The fruit fly memory seems to work in a remarkably similar
way to our own. Like flies, our memories are initially short lived,
but they can become more long term with repeated "training,"
provided that we have suitable rest intervals in between. These
rest intervals are crucial. Anyone who has studied for an exam by
cramming—which is effectively repeated training without rest—
will know that your memory can work well for the day of the
exam, but within a few days your knowledge has completely
evaporated. Fruit flies are no different. Give them repeated training without rest and their memories never stick.

Seymour Benzer's approach to animal behavior was a complete departure from what had gone before. Traditionally, studies of animal behavior had been the domain of naturalists and field biologists. Instinctive behaviors such as courtship, feeding, and fighting were meticulously observed and painstakingly described. Typically, a complex behavior would be atomized into a series of distinct steps. Learning and memory, for example, could be broken down into the acquisition of short-, medium-, and long-term memory. These "atoms of behavior" were made, primarily, in the interests of simplification and descriptive convenience. Naturalists had little interest in how they might relate to the action of genes and molecules. But Benzer wanted to see whether he could map the individual steps of a behavioral pathway to single genes and discrete molecular events. He wanted to dissect learning and memory, and reveal its genetic innards.

In effect, Benzer wanted to do with behavior what Nüsslein-Volhard and Weischaus did with embryonic development. And to begin with, he went about it in pretty much the same way. Benzer bombarded flies with chemical mutagens to create a vast repertoire of fruit fly mutants. He then sorted through these mutants until he found flies that struggled to distinguish the odors in the learning and memory test.

The *dunce* fly was the first learning mutant to materialize. Though physically indistinguishable from a normal fly, *dunce* was the epitome of fruit fly stupidity and quite incapable of learning anything. You could increase the training, or you could make the teaching conditions in the electrified tube more severe by turning up the voltage or increasing the strength of the odors. But it didn't make a blind bit of difference: *dunce* always came bottom of the class.

Benzer hoped that *dunce* would herald the arrival of a deluge of learning mutants. It was obviously a great start, but to resolve a behavioral pathway he needed to find mutant flies with varying degrees of stupidity, flies that might remember their training for a few minutes, a few hours, and so on. But identifying behavioral mutants turned out to be a tricky business. Unlike developmental mutants, many of the behavioral mutants revealed no physical clues to their identity. The only way to identify them was by rigorous training and testing on the olfactory exercise. Consequently, the discovery of new mutants came more as a trickle than a deluge. After *dunce* came *amnesiac, radish, cabbage, turnip,* and *linotte*. Together, they formed a veritable gaggle of fruit fly idiots.

But difficulties remained. The new mutants were all patently stupid, but beyond that they were difficult to distinguish. The problem was that the training equipment was not sensitive enough to quantify intermediate levels of stupidity. Even if there were differences between the mutants, there was far too much experimental noise to distinguish them. Flies were too easily distracted during training, and it was difficult to ensure that all flies were experiencing the same strength of odor and electric shock.

By the 1980s, Benzer had moved on to other things. But Tim Tully, who had connections with Benzer's original research group, decided that he was going to design and build a new teaching machine. This time, nothing was left to chance. He built a special plastic training tube with an electrifiable grid that ran evenly throughout the floor. He used vacuums to ensure that the flow of odor over the flies was smooth and constant. And he did everything possible to minimize outside influences, even providing the flies with an elevator that transferred them from the training area to the choice point of the T-shaped maze.

Tully spent four years designing and building his new teaching machine. With wholesale improvements to the experimental setup, the flies now had little excuse for their minds to wander. Freed from distractions, they could concentrate on the learning task in hand.

With far greater resolving power than the previous design, Tully's machine turned out to be a huge success. Now, many of the learning and memory mutants were easy to distinguish. A *linotte* mutant (the name comes from the French *tête de linotte*, which roughly translates as "birdbrain") was, for example, identified as a fly that has problems remembering what it has learned in the first three hours after training.

The *linotte* fly stood out from the mutant crowd in more ways than one. Its defective *linotte* gene was the product of a novel form of mutagenesis, one that involved neither chemicals nor X rays. By the 1980s, these traditional techniques had become old hat. For any fashionable fruit fly biologist, jumping genes were now the way to manufacture new mutants.

Jumping genes—known as "transposable elements" in learned circles—are genetic parasites. They are short stretches of DNA that "live" and multiply within the DNA of a host organism—a bacterium, a fruit fly, even a human being. There's no escaping them; these selfish pieces of semilife have been littering chromosomes for millions of years, and are a part of the genetic furniture.

A jumping gene's one and only pastime is to play hopscotch along the chromosomes, inserting itself into the host's DNA willy-nilly, with little concern for the consequences of its actions. For the host, these mobile elements are a genetic liability. When a jumping gene inserts itself into one of the host's own genes, it can cause some serious mutational damage.

In the 1980s, fruit fly biologists realized that if they could harness the power of these troublemakers, jumping genes held huge experimental potential. Not only did they offer a novel way of inducing mutations, they also represented a perfect delivery system for putting new, foreign genes into flies.

Jumping genes come in a variety of different flavors. But the most celebrated of the fruit fly transposable elements, and the first type to be isolated and purified, was the P element, discovered in a wild population of *Drosophila melanogaster* in the 1950s. Nobody is quite sure how it got there, because the element has not been detected in any closely related *Drosophila* species. One possibility is that the element came from a distant relative, via a virus or bacterium. Whatever the mode of transmission, once the P element was established, it spread like wildfire. By 1980, P elements could be found in *Drosophila melanogaster* populations worldwide.

The P element has lived up to all experimental expectations. Without a hint of hype, you could say that it has revolutionized fruit fly molecular biology. It can be injected directly into fruit fly embryos, where it can wreak mutational havoc, cluster-bomb style, on the fly's DNA. Alternatively, extra DNA can be tagged on the end of the P element before it is injected. In this form, it becomes a molecular envelope, delivering a wide variety of genetic messages to the fruit fly's DNA.

The P element's versatility has made it an invaluable tool for dissecting the genetics of fruit fly learning and memory. It is perfect for the simple genetic arithmetic of adding and subtracting genes, creating engineered flies with a missing gene here or an extra gene there. And the performance of these engineered flies in their learning and memory exams paints a revealing picture of how genes conspire to commit learned experiences to memory.

In the late 1990s, Tim Tully took some normal fruit fly embryos and, with a fancy bit of genetic engineering, knocked out their *linotte* gene. His genetic tinkering did not stop there, however. Tully also mailed a P element envelope containing a normal copy of the *linotte* gene and a so-called heat-shock promoter tagged on the end of it. The heat-shock promoter acted as a control switch for its neighbor. At room temperature, the mailed *linotte* gene would remain permanently switched off; but if the heat was turned up on the engineered flies, to a temperature of about 95 degrees Fahrenheit, the gene would be switched to the "on" position and start making its protein.

All the engineered flies were reared at room temperature. When the adult flies were trained and tested, they all displayed learning difficulties that typified a defective *linotte* gene. But then Tully performed a little piece of genetic magic. He dunked a bottle of the engineered flies into a hot bath. As the heat percolated through the fruit fly bodies, normal *linotte* genes everywhere swung into action. Inside each engineered fly, the mailed *linotte* genes that had so far lain dormant were suddenly switched on.

Three hours later, the flies were put through their paces once more: the simple training exercise, followed by the choice test in the T-shaped maze. Would the flies be as dimwitted as ever, or would they display a newfound intellect? The results were unequivocal. Pass rates were back up to the 90 percent mark, the level you would expect from a batch of normal flies. The flick of a genetic switch had rekindled the learning powers of the flies. It was a remarkable transformation, and the first time that gene therapy had been used to treat a learning disability.

The *linotte* gene is not the only gene that works as an on/off switch in fruit fly learning and memory. The *CREB* gene, for instance, seems to work in a similar fashion. It turns on long-term memory in flies that have received a spaced sequence of training exercises; turn off or mutate the *CREB* gene, and flies never acquire a long-term memory, no matter how much training they receive. But turn it back on again, via a heat-shock promoter, and the fly's long-term memory potential is miraculously revived.

The *CREB* gene can do more for a fly than simply return its memory banks to normal, as Tully discovered when he engineered flies with extra copies. An additional dose of the *CREB* gene resulted in flies with photographic memories. The flies no longer required repeated, spaced training exercises to acquire long-term memory. They learned in one lesson what it took normal flies ten lessons to learn. One simple training exercise is all it took to make a fruit fly mastermind.

Genes such as *dunce, linotte,* and *CREB* have vindicated Seymour Benzer's original vision. They are living proof that behavior is amenable to genetic dissection. Slowly but surely, the atoms of behavior are being translated into the atoms of inheritance.

Identifying genes is one thing. But understanding how these genes work, how their instructions are translated into molecular events and memories, is a different matter altogether. More than twenty genes associated with learning and memorizing odors have been identified so far, but the function of many of them remains obscure. Nevertheless, the genes that have been decoded are already beginning to paint a tantalizing picture.

Memorizing new information is associated with physical and physiological changes to the nervous system. In the case of a single training exercise, these changes—and the memories that result—are transient and short lived. Only repetitive training produces a more permanent increase in both the number of neural connections and their sensitivity.

Genes like *dunce, linotte,* and *CREB* form part of a biochemical pathway inside cells that coordinates these physical changes. These genes—or, more accurately, their protein products—translate electrical impulses received at the cell surface into physical changes in the nerve cell itself.

The details of the learning and memory pathway are full of molecular twists and turns. But viewed from a safe distance, the pathway reveals a remarkable feature. There seems to be a striking correspondence between the stage of the pathway and the phase of learning. The *linotte* gene, for example, codes for a protein that acts early on in the pathway, and affects an early phase of learning. In contrast, the *CREB* gene is associated with the terminal end of the pathway and is crucial for long-term memory.

One way to visualize this pathway is to think of the series of biochemical steps as stepping-stones that connect two sides of a riverbank. One side of the bank represents the starting point, a state of complete ignorance; the other side represents the end point, the acquisition of long-term memory. Each stepping-stone is a stage in the learning and memory pathway. A mutation in one of the genes in the pathway would be like removing one of the stepping-stones from the river. Learning and memory could then progress to the phase just prior to the missing stone, but no further.

This view is certainly simplistic. Undoubtedly, there is still much to learn about learning and memory. But fruit flies have already created a sensation. Learning and memory have been translated into a series of biochemical switches that can be manipulated by simple genetic additions and subtractions. It's enough to make an antireductionist wince, but the question has to be asked: Could the same be true in humans?

Learning and memory genes similar to those found in flies have already been discovered in other animals. Humans, mice, rats, nematode worms, and sea slugs all have a gene whose DNA sequence matches that of the fruit fly *CREB* gene. In fact, *CREB* seems to be a universal molecular switch found throughout the animal kingdom. In mice, for instance, *CREB* turns on long-term memory, just as it does in flies. Disrupt the *CREB* gene and you create a breed of mice with nothing but short-term memories on their mind. Like the genes for development, learning and memory genes look like a conservative bunch.

If we are working to the same genetic template as fruit flies, then it points to an exciting and perhaps frightening future of memory manipulation. Can you imagine climbing into a hot bath to learn about Einstein's theory of relativity, and then forgetting it again as soon as you step out? And how about engineering CIA agents with an extra dose of a *CREB* gene to give them a photographic memory? With the heat-shock promoter included, spying would have to be restricted to hot spots such as Africa and the Middle East. But with the heat-shock promoter taken out, espionage could become truly international. Then again, without some element of temperature control, your brain might soon become overburdened with information.

On a more serious and practical note, fruit flies suggest a future in which new drugs and gene therapies are used to treat congenital learning disabilities, stroke victims, and those suffering from conditions such as Alzheimer's disease. Memories lost by blows to the head could be resurrected, while painful and traumatic memories could be chemically excised.

It almost sounds too good to be true. And perhaps it is just another overhyped and soon-to-be-forgotten vision of the future. Only time will tell. But reread the above paragraph ten times (with appropriate rest periods in between) and it just might stick.

The fruit fly's ability to learn and memorize smells is not just some kind of eccentric party trick. There are good evolutionary reasons why flies have acquired the capacity to memorize odors. Flies rely on olfactory cues to navigate their way around their miniature world. Without the ability to distinguish the "good" smells of food, mates, and egg-laying sites from the "bad" smells of danger, fruit fly life would be a frustrating—and very quickly fatal—occupation.

Undoubtedly, one of the most important odors in the fruit fly memory banks is alcohol. Alcohol is a by-product of rotting and fermenting fruit. Because it is a highly volatile organic compound, a piece of ripe fruit will ooze an alcoholic vapor trail. An ability to sniff out the trail means that flies can make relatively light work of locating their favorite egg-laying and feeding sites.

Indeed, the fly's evolved sensitivity to alcohol may offer an insight into the evolutionary origins of our own love affair with the demon drink. Any animal that depends on ripe and rotting fruits would benefit from having antennae, or a nose, sensitive to alcohol. Millions of years ago, our hirsute ancestors were living on ripe fruits in the forest canopy. Because these fruits would have been difficult to find among a maze of green leaves, an ability to sniff out an alcoholic vapor trail would have provided a distinct advantage.

Could our predilection for alcohol be an evolutionary hangover from the past? Each time we walk into a bar, are we echoing our ancestors' search for a decent bite to eat? There is certainly no shortage of circumstantial evidence to back up the idea. Consider our drinking habits, for example. We tend to drink alcohol in fairly dilute form—beer and wine. When we have stronger drinks such as spirits, we usually water them down. Out in the wild, amidst the forest canopy, fermentation also results in fairly dilute concentrations of alcohol. Even at their metabolic peak, yeasts can only make fruits alcoholic to the tune of 10 to 15 percent. Is it merely coincidence that the concentrations of alcohol we prefer are the same ones that our ancestors would have encountered in the wild?

And to add fuel to the speculative fire, why is it that we, like fruit flies, tend to live healthier and longer lives with a regular, if modest, intake of alcohol? If low concentrations of alcohol were an everyday part of our ancestors' diet, then perhaps we should not be too surprised if evolution has attuned our physiology toward it.

Of course, drinking in moderation is not always easy to control. Any animal that feeds on ripe and rotting fruit is faced with the possibility that an afternoon's feeding could inadvertently turn into an evening on the tiles. For fruit flies, butterflies, monkeys, elephants, and millions of other fruit-eaters, getting drunk is an occupational hazard, an inevitable consequence of feeding on a particularly ripe batch of fruit.

When it comes to the intoxicating effects of alcohol, we share some uncanny behavioral similarities with flies. In fact, there are three phases of fruit fly drunkenness, which should be familiar to us all. Initially, there is a euphoric, boisterous phase, in which the flies are twitchy and hyperactive. This is when the fly would begin to lose its inhibitions, if it actually had any to start with. Next comes the uncoordinated phase, when the fly struggles to walk in a straight line. Flying might be possible, but it's just not worth the effort. And finally, there is the collapse-into-a-coma phase, when the fly loses consciousness, only to wake up in the gutter, or, worse, a predator's stomach.

The similarities do not end there. The fly's tolerance to alcohol is remarkably similar to our own. Flies tend to become inebriated when their blood alcohol concentration reaches about 0.2 percent (0.2 grams of alcohol per 100 milliliters of blood). Compare that with the legal limit for driving in most countries—around 0.1 percent. With so much in common, is it any wonder that flies are being touted as models for the study of human alcohol abuse and addiction?

For our ancestors, a boozy afternoon was probably an occasional and unexpected bonus, providing some light relief from the mundane routines of tree-top living. But now that we are down on terra firma, in an environment where alcohol is always available in excess, things can only too easily get out of hand. Alcoholism may be one instance where our evolutionary heritage gets the better of us.

Of course, we are not programmed to become alcoholics. There are probably numerous factors, both biological and social, that underlie the onset of alcoholism. But there's no doubt that heritable factors can play an important part in the disease. Genes go some way to explaining why alcoholism tends to run in families, and why individuals vary enormously in their physiological sensitivity to alcohol.

For some people, one pint of lager may be enough to leave them legless. Others may require pints in double figures to achieve the same effect. Likewise, fruit flies vary widely in their sensitivity toward alcohol. Genes mean that some flies can hold their booze much better than others. In fact, there is some evidence to suggest that flies that live in particularly alcoholic environments—such as vineyards and wineries—have evolved a much higher tolerance to alcohol than those from elsewhere.

In humans, alcohol tolerance seems to go hand in hand with alcoholism. People who are more resistant to the intoxicating effects of alcohol are more likely to become alcoholics. So identifying the genetic and molecular basis of alcohol sensitivity has been seen as the first step toward developing suitable treatments and cures for alcoholism. The fly might seem an unlikely candidate for such a role, but, as its involvement in the study of learning and memory has shown, an organism doesn't need complexity to make a meaningful contribution to genetic studies of behavior.

But how do you identify genes for alcohol sensitivity? How do you accurately distinguish flies that can hold their drink from those that can't? The answer is an ingenious piece of apparatus called an "inebriometer." The inebriometer is essentially a meter-tall vertical glass column filled with alcohol vapor. Inside the column, perforated sloping platforms are fixed at various heights, giving the flies a place to sit and preen themselves.

To begin a "run," about a hundred flies are introduced to the top of the column. As the alcohol begins to take effect the flies lose their balance, and roll off the platforms down to the next level below. When a fly is totally incapacitated, it is funneled by the platforms out of the column at the bottom. The average time taken for the flies to emerge, dazed and confused, from the base of the column is used as a reliable indicator of their sensitivity to alcohol.

Inside the inebriometer, normal flies last for an average of twenty minutes before they fall out of the column. But in 1998, a mutant fly materialized that seriously struggled to hold its drink. Instead of the normal twenty minutes, this mutant strain of fly was falling out of the inebriometer after fifteen minutes. There would be no danger of breaking the bank on a night out with this fly; hence the mutant's name—*cheapdate.*

The date might turn out to be a cheap one, but it's unlikely to be intellectually stimulating, because *cheapdate,* it transpires, is not a new gene but a new mutant version of *amnesiac,* the learning and memory gene. In other words, *amnesiac,* a gene that affects the ability to learn, also affects alcohol tolerance. This genetic coincidence is not unique. Many of the learning and memory mutants have turned out to have an increased sensitivity to alcohol. Stick them in the inebriometer and they always fall out quicker than the rest.

There is still much dissecting to be done before the genetics of alcohol-induced behaviors makes any kind of molecular sense. But we already know that the genetics of alcohol tolerance overlaps the genetics of learning and memory. Genetic engineering can be used to "rescue" *cheapdate* mutants in the same way it was used to rescue *linotte* and *CREB* mutants. A *cheapdate* mutant can be turned into a more alcohol-tolerant fly with the simple flick of a genetic switch. It's still too soon to tell, of course, but the fly may yet find solutions to the problem of our love-hate relationship with alcohol.

After a hard day on the teaching machine, or an afternoon playing "drink till you drop" in the inebriometer, there is nothing a fly likes more than settling down for a nice nap. But even at rest, the fly is not free from behavioral interrogation. Few animals have had their patterns of sleeping and waking more heavily scrutinized than the fly.

Fruit fly sleep is slightly different from our own, but it amounts to the same thing. Being insects, flies have no eyelids to shut when they are feeling tired. Nevertheless, when it gets dark, they slip into soporific mode. The night is a time for rest and relaxation, a period in which to recover from the frantic pace of the day.

The fly lives its life to an approximate twenty-four-hour rhythm, just as we do. Its body is attuned to the regular cycle of night and day. It wakes up in the morning and it goes to sleep in the evening. Little wonder, then, that when it comes to unraveling the molecular mystery of biological clocks, the fly has been much in demand.

The world first woke up to genetic studies of biological rhythms in the early 1970s, when Seymour Benzer and his student Ronald Konopka announced the discovery of the *period* gene. Although nobody had a clue how it worked, the gene seemed to function as a timekeeper for the fly's twenty-four-hour day. Mutations in the *period* gene produced flies with aberrant biological rhythms. One mutant had its clock running too fast and lived its life on a nineteen-hour cycle. Another was on a slow clock—a twenty-nine-hour cycle. And a third mutant seemed to have lost its sense of rhythm altogether, sleeping and waking at random.

When it came to timekeeping, the *period* gene had its fingers in many behavioral pies. Not only did it regulate the fly's twenty-four-hour clock, it also controlled the rhythm of the fruit fly's wing beat during courtship. The *period* gene was like an alarm clock and a metronome rolled into one.

While there was no shortage of discoveries showing what the *period* gene could do for a fly, molecular explanations of how a gene, and the protein it encodes, could manage to keep time were a bit thin on the ground. It was not until the late 1980s that the first hint of the clock's internal mechanism sprang into view.

The evidence emerged from a particularly gruesome experiment that saw hundreds of flies lose their heads. Not since the guillotine madness of the French Revolution had the world seen ritual decapitation on such a grand scale. At hourly intervals throughout the day, flies were removed from their bottles and their heads were removed from their bodies.

It might not have been pretty, but the outcome of all this execution was an hour-by-hour snapshot of the chemical goings-on of the *period* gene. Chemical analysis of the fruit fly heads revealed that the concentration of the *period* protein in the fly's brain fluctuated throughout the day. The protein reached a peak during the hours of darkness and then declined to a low point during the afternoon.

In 1994, a second timekeeping gene, *timeless,* was discovered. Any flies unfortunate enough to have a mutant version of the *timeless* gene had trouble getting any sustained sleep. While their normal laboratory comrades were tucked up in bed for the night, *timeless* mutants were still actively pacing the glass, no doubt desperate to have a nap but biochemically incapable of doing so. Like the *period* protein, the concentration of the *timeless* protein in the fruit fly brain was found to fluctuate around the clock, reaching a peak in the night and dipping to a low point around mid-afternoon.

The *period* and *timeless* proteins—Per and Tim—have turned out to be partners in timekeeping. The two proteins are like the two halves of an amorous couple. Stick them both in a Petri dish and they join together in a molecular embrace.

Inside the brain cells of the fly, this partnership never lasts. Like fickle lovers, the proteins pair up, then separate, then pair up again. It's an endless cycle, a continual on-off romance. On-off, on-off, tick-tock, tick-tock; each swing of the pendulum means another day in the life of a fly. It is the pace of this constant molecular marriage and divorce that seems to set the fly's internal rhythm.

To put things in a bit more detail, imagine that we are sitting inside one of the fruit fly's brain cells. We are at the heart of the action. It's about five hours before dawn and the fly is resting. Inside the cell, there are Per-Tim couples everywhere. At this time of the day, their concentration is at its peak. The Per-Tim couples are making their way into the nucleus of the cell—the home of the genes. The nucleus has a strict door policy: it allows Per and Tim proteins inside only if they are in a couple. Protein singles are denied entry. Once inside the nucleus, the protein couples begin turning off the *period* and *timeless* genes. In effect, the Per-Tim partnership regulates its own production. Turning off the *period* and *timeless* genes cuts the production of Per and Tim and blocks any further accumulation of Per-Tim couples.

The changing levels of Per and Tim have a direct effect on the production of natural sedatives, like melatonin, within the fruit fly brain. Declining levels of Per and Tim mean declining levels of sedative. By dawn, the fly is wide awake and ready to start a brand-new day.

Dawn, and daylight, bring further changes. While Per is totally at ease with the cold light of day, Tim can't stand it. In fact, Tim is a werewolf of the protein world. Stick Tim in the light and it starts to lose its molecular integrity and fall apart. By necessity, Tim is very much a night owl.

So dawn brings with it the breakup of the Per-Tim partnerships. Gradually, these protein pairs disintegrate and disappear from the cell nucleus. As they diminish, so does their controlling influence on the *period* and *timeless* genes. These genes have sat dormant since the middle of the night. But by noon, they are up and running again.

Because it is still daylight, there wouldn't be much point making fully formed Tim protein. So the two proteins are stockpiled in precursor form. Only when the lights go out, after sundown, does the production of Per and Tim proteins go into overdrive. The rising levels of protein switch on sedatives in the fruit fly brain, and the fly is happy to call it a day.

But while the fly rests, the two proteins continue to build up in the brain cell. Once they have reached a critical concentration, partnerships will be reestablished. The Per and Tim proteins continue to take one another by the hand until the concentration of couples has reached a peak, at about five hours before dawn, and we have completed our cycle.

By necessity, this story of the circadian cycle is part fact and part fantasy. There is still much to learn about what makes the fly's clock tick. If the time of day represents the degree of understanding, then biologists are still fumbling around in the predawn darkness. No doubt many more fly heads will roll before a more complete picture emerges.

Some things, however, are clear. Genes and gene products may be the cogs and gears that keep the clock ticking, but it is daylight that sets these molecular controls. Daylight decides when the Tim protein can accumulate, and when the Per-Tim partnerships degrade.

This sensitivity to daylight means that the fly's circadian rhythm, like our own, can adapt to changes in time zones. You can demonstrate this effect quite easily by manipulating the fruit fly's exposure to light. If you give flies an extra hour of daylight after dusk—around 10 P.M., for example—then you delay the accumulation of Tim protein in the fruit fly brain. The result? The fly's circadian rhythm is postponed and reset by about four or five hours. This is what happens to our own internal clocks when we travel west on long-haul flights and experience a lengthened day.

In contrast, if you expose flies to an hour of light before dawn, then the Per-Tim partnerships degrade earlier than expected, and the fly's behavioral rhythm is advanced by a few hours. Again, our own internal clocks readjust in the same way when we travel east and experience a shortened day.

Whether flies also experience jet lag while their body clocks readjust is a matter still open to speculation. But it wouldn't be too surprising to find out that they did. Flies have already shown themselves to be consummate role models for biology. As their genes bring scientific enlightenment to human behaviors, it adds yet another feather in a cap already bristling with plumes.

Despite its small size and dubious habits, the fly is a sophisticated little creature. It can learn and memorize information in ways that echo our own education. It can indulge itself in booze—with all-too-familiar consequences. And, like us, it can retire to bed for a good night's sleep when things get too much, to wake at dawn, bright-eyed and bushy-tailed. While the fly may have sex on the brain, there still seems to be space for a lot more besides.

So, given a choice between a domestic dog and a domestic fruit fly, which would you rather have as a pet? Most people would probably plump for the dog. Dogs are soft and cuddly; each one is an individual with its own unique personality. Fruit flies, on the other hand, are a different kettle of fish. It's always going to be difficult to form a long-lasting pair bond with an animal that drops dead before you've had time to think up a suitable pet name.

Dogs may indeed have the competitive edge. But at least you can teach an old fly new tricks.

5

THE SINISTER SIDE OF SEX

I'm standing on the curb of one of the busiest roads in London. A torrent of traffic, six lanes deep, forms an intimidating obstacle between me and the pavement on the other side of the road. I watch as countless other pedestrians approach the curb and then turn back, too fearful to attempt the crossing.

I'm on my way to University College London, to meet the fly. It's a casual visit, a chance to see what a modern fruit fly laboratory looks like, and to gain a glimpse into how it is run. But I'm also after some more specific fruit fly information. Today, I'm hoping to get the latest news on the fruit fly's sinister sex life.

I turn off the main road and walk down a gloomy cobbled street. The building at the end of the row, a dumb juxtaposition of brick and glass, looks as if it were designed by the sort of failed 1960s architect that was all a poverty-stricken university could afford. I wipe my feet, go inside, and take the lift to the top floor.

As the lift doors open I'm hit by a peculiar smell. It is the smell of a brewery whose levels of hygiene have dropped below acceptable standards. Yeast dominates the stench, but a host of lesser-known odors lurk in the mix and combine to produce an uncomfortable olfactory blend.

Out in the hall, there are crates of half-pint milk bottles stacked up against the walls. Memories come flooding back of school days long ago, when a half-pint of tepid milk was all part of a young child's education. But a growing sense of nostalgia is nipped in the bud when I notice that the bottles are smeared with fruit fly excrement. I move on, passing through a pair of heavy double doors.

Suddenly, I'm in a completely different world. Everything is white, shiny, modern, and new. The contrast with the building's depressing exterior could not be more extreme. The whole place oozes big-grant opulence. Clearly, the science is going well here.

The corridor is narrow, not much more than a meter wide. On either side, sliding doors mark the entrances to constant-temperature rooms—windowless chambers in which the temperature and light can be set for maximum fruit fly productivity. Out in the wild, the vicissitudes of the weather can play havoc with a fruit fly's egg-laying ambitions. But stick them in a room at a constant 77 degrees Fahrenheit, with twelve hours of light and twelve hours of darkness, and as sure as eggs is eggs, the flies will not let you down.

One of the doors is open and I take a peek inside. The room is tiny, the size of your average garden shed, and made to feel even more cramped by the floor-to-ceiling shelving that crowds the walls. The shelves are covered in more half-pint milk bottles, hundreds of them, each one housing hundreds of flies. There could be as many as a million flies in all.

Also inside the room is the fruit fly biologist Tracey Chapman, my host for the day. Tracey has spent most of her research career studying the sex lives of fruit flies. Her speciality is seminal fluid. It is a strange topic, you might think, to devote one's life to. But in fruit flies, there is a lot more to seminal fluid than meets the eye. Fruit fly semen has a touch of the devil inside.

Tracey is sitting at a desk, staring at a floodlit, transparent sandwich box, which has been turned into an impromptu arena for fruit fly courtship. I step into the room and squeeze myself into a seat to get a good view of the action.

Inside the box, hundreds of flies are pacing the plastic. Many of the flies have already paired off, the female taking the lead, the male following close behind, his eyes seemingly fixated on her creamy-white abdomen, bloated with eggs.

The males seem to be going about their business with an urgency that is completely lacking in the females. You can almost sense the sexual tension in their jerky, neurotic movements. One male follows a female around the box until a single female walks past in the opposite direction. Confused by a conflict of desires, the male pauses to consider his options. But by the time he has made up his mind, both females have disappeared into the crowd, and he is left alone in a deserted patch of plastic.

For such small and supposedly simple animals, fruit fly courtship is a remarkably elaborate affair. True, the fruit fly mating game has none of the head-banging brutality of the elephant seal, or the regal pomp of the peacock. But fruit fly courtship does have its own peculiar charms. After all, what other male counts on a curious mix of cunnilingus and song to illustrate the integrity of his semen?

A male tails a female and I watch as his chat-up routine unfolds. His right wing, held perpendicular to his body, is vibrating furiously. He continues for a few seconds and then switches to the left wing. Then back to the right again. Occasionally, he sticks both wings out and vibrates in stereo.

Wing vibration is the male's way of singing a song. Without amplification, the song is more or less inaudible to the human ear. With amplification, biologists have discovered that the song is all rhythm and no melody. Wing vibrations produce pulses of sound—beats, if you like—separated by intervals measured in milliseconds. As the song progresses, the male varies the tempo in a cyclical fashion, speeding up, slowing down, then speeding up again. Think of someone playing with the throttle of a two-stroke engine and you should get the general idea:

Pup......Pup.....Pup....Pup...Pup..Pup.PupPupPup.Pup. .Pup...Pup....Pup....Pup......Pup....

The song is supposed to put the female in a romantic frame of mind. But, in this instance, it doesn't seem to be working. It is the female, in fact, who seems to be dictating the terms of their encounter. If she stops, the male stops, too. If he encroaches on her personal space, she extrudes a long spike from her rear end, which she uses to fence him off. The spike is actually her ovipositor—her egg-laying tube—but it doubles up nicely as an offensive weapon.

Every now and again, the male ignores the formal rules of engagement and makes a lunge toward the female's rear end, giving it a quick and surreptitious little lick. If she anticipates his move, she can head him off with her lancet, or tuck her naughty bits underneath her body, where the male can't reach them.

But this time, it looks as though she is going to receive his advances. She stops. The male has his penis poised for action. But then she moves off again, seemingly disappointed by what she has seen. This is, perhaps, no great surprise. The fruit fly penis is a quite pathetic little piece of anatomy, a tiny spikelet that barely protrudes from the tip of the abdomen.

And so it goes on. More singing, more kissing, sometimes a little bit of fondling, as the male gets frisky with his forelegs. More rejections. Stop. Start. Stop. Turn right. It is almost too exhausting to watch. He has been playing a frustrating game of follow-the-leader for over fifteen minutes now. In human terms that equates to about thirty continuous days of courtship without food or sleep. And sex still seems so far away.

Suddenly, the situation changes. Whether through boredom, sympathy, or genuine attraction, the female decides that the time is right. She offers her rear end and the male willingly climbs on board. He arches his abdomen underneath his body so that his penis can engage with the female's vagina. Here, the penis comes into its own. It may be ridiculously small, but a male fly needs his penis like a tent needs a peg. Together with his forelegs and a pair of genital claspers, the penis anchors him securely to the female.

The two flies shuffle about a bit, making last-minute adjustments to their genitals. Fruit fly copulation normally lasts about twenty minutes. But today, these flies will not be so lucky. Within a minute or so of their mounting, Tracey has vacuumed up the pair with a piece of rubber tubing. The flies are sucked out of the box and into a small glass tube. They ricochet around the walls for a few seconds before finally coming to rest on opposite sides of the tube. Coitus doesn't get much more interruptus than this.

The flies look disappointed, and so am I. But we are not here for the cheap thrills of a fruit fly peepshow. Separation is all part of the experimental script. Tracey wants to identify the females that are eager to mate—but at the same time, she doesn't want them to be inseminated. So she has to separate the pair within two minutes of mounting, while the male is still fiddling around with his genitals.

Tracey is interested in the hidden dangers of seminal fluid, and the strategies that females use to avoid them. The experiment is all part of ongoing research into the dark side of fruit fly sex, research that has seen the image of seminal fluid undergo a transformation. No longer can it be considered a harmless medium for the transport of sperm. Fruit fly seminal fluid is a malevolent cocktail of chemical weapons in a never-ending sex war.

Sex has always been a bit of a sticky issue. Evolution, in its blind and merciless way, has shaped a world in which males and females live in perennial conflict over issues of parenthood. Both sexes want the same thing—to sire as many healthy offspring as possible—but basic biological differences mean that they go about it in different and conflicting ways.

In simple terms, the conflict boils down to gametes. Males produce millions of sperm, whereas females produce much smaller numbers of eggs. This discrepancy in gamete production means that a male can fertilize many more eggs than a single female can produce. From the male perspective, promiscuity is the best way to maximize reproductive success. With all males pursuing a policy of sexual greed, there will inevitably be both winners and losers in the competition for mates.

Females see sex from a completely different angle. Since egg production places a limit on the number of offspring they can produce, the best way of maximizing reproductive potential is to be choosy about whom you mate with, or how often you mate. With so much sperm on the market, it pays to shop around and get the best deal for your eggs. Choosing the sperm of a fit and healthy male, rather than a scrawny ne'er-do-well, will pay long-term evolutionary dividends.

In short, the conflict of interest between males and females is a conflict of quantity versus quality. Darwin recognized this asymmetry that lies at the heart of sex. It provided him with an explanation of why the males of many species are so much larger and more gaudy than their mates.

Where there is competition among one sex—usually, but not always, the males—for the other, any feature or behavior that increases the chances of parenthood will be favored by what Darwin called "sexual selection." Unlike natural selection, sexual selection depends on the struggle for mates rather than the struggle for existence. Darwin believed that antlers of red deer, for example, or the elaborate plumage of the male peacock, evolved not because they increased a male's chances of survival, but because they increased the chances of fathering offspring.

In short, evolution has turned males into a nation of bullies and showmen. In some species, the males will struggle for mates by trying to beat their opponents into submission. In others, they will broadcast their rude state of health with shameless displays of exhibitionism.

But while many species prefer to fight their sexual battles out in the open, others adopt a more stealthy approach. In insect species, for example, the struggle for paternity sometimes extends inside the female. During copulation, males can displace the sperm of rivals before depositing their own.

To assist them in their seminal spring-cleaning, evolution has blessed the insect penis with all manner of fancy gadgets. The penis of the Mediterranean rabbit flea, for example, with its hooks, levers, barbs, and springs, has more accessories than your average Swiss Army Knife, and is widely regarded, by those in the know, as the most complicated penis in the world. Some penises are glorified scoops, others have whips and tassels. One species of dragonfly even has an organ that can be inflated once inside the female, forcing rival sperm out into the fringes.

There seems to be no limit to the lengths that some male insects will go to guarantee paternity. Many species cement over the female's genital tract after mating to prevent other males from mating with her. Some species also "mate" with males in an attempt to decommission the genitals of their rivals. Even more bizarre is *Xylochoris maculipennis,* a bug that has abandoned all notions of sexual etiquette. Instead of inserting his penis in the usual place, the male uses it like a hypodermic syringe, injecting his sperm through the female's body wall. The sperm then swim around inside her body until they bump into her eggs. Even more deviously, males inject their sperm into the bodies of rivals. Once inside, the sperm swim to the testes, where they wait to be ejaculated.

In the competition for mates, *Xylochoris maculipennis* seems to have evolved an extreme "fire and forget" strategy, in which the sperm do most of the work. But the fruit fly extends the sex war even further. Battles are won and lost not with penises—the fly is plainly ill-equipped for any genital jiggery-pokery—or even sperm, but with proteins in the seminal fluid. Every time a male inseminates a female, he delivers a cocktail of drugs designed to wrestle control of the female's body and mind, and make her act in his own interest.

These new revelations have challenged the traditional pipe-and-slippers image of seminal fluid. Conventional wisdom has it that seminal fluid is simply a liquid medium for the transport of sperm. Its varied assortment of chemicals are there—so the story goes—as a sort of chemical packed lunch, to assist the sperm on its long journey in search of eggs. But this remains largely speculative. With the exception of the seminal fluid of fruit flies and a few other insect species, nobody really has a clue what most of the chemicals in seminal fluid actually do. Even in humans, the exact function of most seminal constituents remains obscure.

The first big hint that fruit fly semen was not all sweetness and light came in the 1950s, when insect physiologists discovered that semen could manipulate a female's behavior. Injected directly into females, the seminal fluid suppressed libido and induced egg-laying. If the last male to mate could deliver these kinds of effects, along with his sperm, then it was clearly going to boost his chances of paternity.

Years later, the chemicals responsible for these effects were traced to the so-called accessory glands, which sit next door to the testes. Inside the glands, proteins are manufactured and stored in preparation for ejaculation, and their launch into enemy territory. About twenty proteins have been identified so far, but current estimates suggest that as many as a hundred might be mixed up in the molecular offensive.

The struggle for parenthood sees seminal proteins reach out into all corners of the female's body. Some stay close to home in the genital tract, others move further afield, traveling via the bloodstream to exert their influence on the brain. In the clamor to get genes into the next generation, it seems that evolution has unwittingly turned the female's body into a war zone, with battles being fought on all fronts.

It is difficult to relate this savage tale of molecular warfare to the picture of the courting flies in front of me. Violence seems to be the last thing on their minds. Those that are still left in the arena have retired to the perimeter of the sandwich box, seemingly exhausted by the morning's activities. Two hours on a small chair has left my backside a bit the worse for wear, too, so I decide to stretch my legs. A lull in the proceedings is the perfect opportunity to have a nose around the rest of this fine scientific establishment.

I am wandering the corridors, putting my head around doors, making small talk, trying to gauge the atmosphere of the place. In one room, there are about half a dozen people sitting in a neat line at a long bench. Each is hunched over a microscope, cocooned in a private world, counting, measuring, or watching flies. Business is being conducted in near stately silence, save for the murmur of a muted radio in the corner, a small concession to the outside world.

Not everyone here is studying the sex life of fruit flies. There are a number of different research projects on the go. How does diet affect the life span of the fly? Why do fruit flies get larger the farther away from the equator you go? What happens to flies when they become inbred? These and a dozen other questions are keeping people glued to their microscopes.

Despite their different goals, everyone in the laboratory seems united by their industry and production. This is not too surprising. The fruit fly is a demanding little beast. The very thing that makes it so useful as a scientific tool—its productivity—also condemns fruit fly workers to a lifetime of slavish dedication. Just keeping the flies alive, the basic maintenance of feeding, cleaning, and housing, is an immense amount of work in itself. Experiments merely add to the burden.

Fruit fly research is not so much an occupation as an entire way of life. It routinely consumes twelve-hour days and seven-day weeks. Holidays are infrequent or nonexistent. Even public holidays are forfeited if the fruit fly demands it. Little wonder, then, that the fly inspires a mix of love and loathing. It is both the bringer of great scientific riches and the taker-away of a social life. The same animal that can give you data by the yard can also keep you huddled over a microscope when the rest of us are getting ready for bed.

I walk down another hallway, toward the kitchen, the heart of the whole industrial operation. Food fuels the fruit fly lust and keeps the breeder reactor ticking over. While adult flies are opportunist eaters, the maggots, like most youngsters, are much more fussy about their food. They prefer the yeasts that grow on rotting fruit rather than the fruit itself. So today, as on every other day, yeast is on the menu.

In one corner of the kitchen, a cooking pot big enough to hold a small child sits on a flaming gas hob. Inside the pot, a glutinous, pale-brown mixture bubbles away like a hot mud spring. Every now and again, the mix vomits halfhearted geysers into the air. This steaming brew, a pungent mix of yeast, agar, sugar, cornmeal, and water, is what keeps the laboratory fly in business. When it has cooled, the liquid is dispensed into clean milk bottles, where it hardens to form a kind of cake. Few cakes are as versatile as this one. The flies eat it, lay their eggs in it, and eventually collapse and die on it.

The fruit fly kitchen, like all kitchens, is a place for washing up as well as cooking. Thousands of milk bottles pass through every week. Each one must be scrubbed clean and sterilized before it can become home to the next generation of flies. Nobody wants parasites or disease getting a foothold. Yet despite these precautions, infections do emerge. Mites, in particular, are the nightmare that everyone dreads. They can make life a misery for fly and human alike, undoing months of hard labor.

I leave the kitchen and return to the sandwich box, to find the flies in more or less the same position as I left them, sitting alone, along the edges of the box. Maybe it is a sit-down protest, a collective stand against the monotony of their diet. Or maybe it is time for me to look at something else.

On a bench in another room, Tracey is setting up a piece of electrical equipment. Although I don't yet know what it is, I feel uneasy about the device. There is a menacing medical quality about it. Wires connect the business end—a pair of fine tungsten filament electrodes—to a black Bakelite box. It doesn't really have a name, but once Tracey has given me a rundown of what it can do, I decide to call it the "castrometer," for reasons that will soon become clear. Fearful it may be, but, as I discover, the castrometer helped to solve a fruit fly sexual puzzle that dragged on for nearly thirty years, a puzzle whose solution adds a bitter aftertaste to the seminal protein story.

It was back in the 1960s when someone first noticed that sex was bad for a female fruit fly's health. Experiments showed that promiscuous females lived shorter lives than their more abstemious relatives. By itself, this result was not as sinister as it sounds. One possible explanation for the reduction in life span was that female flies were simply following the "live fast, die young" philosophy, trading less time on Earth for more genes in the next generation.

But subsequent experiments proved this not to be the case. Promiscuous females were not only dying younger, they were also laying fewer fertilized eggs. For female flies, the message was simple: too much sex can kill you.

But what aspect of sex was so costly? Was it the stress caused by constant male harassment? Was it physical injury caused by the rough-and-tumble of copulation? Or were sexually transmitted parasites and diseases to blame?

Still another possibility was that seminal fluid was the culprit. If fruit fly semen was mildly toxic to females, repeated inseminations could cause premature death by poisoning. It was an intriguing hypothesis, but one as yet without any evidence to back it up.

Over the next thirty years, fruit fly biologists worked hard to tease apart the various aspects of courtship and copulation and examine how they affected a female's life span. In the late 1980s, Kevin Fowler and Linda Partridge, then both at the University of Edinburgh, now at University College London, discovered that there was something specifically about mating that could shorten a female fruit fly's life.

The key to Fowler and Partridge's success lay in their ability to distinguish experimentally the effects of courtship from those of mating. Their plan was deceptively simple. They would take some normal female flies and divide them into two groups. The first group would be let loose with males that could court and mate normally. The second group would be paired up with males that could court normally but were unable to mate. Fowler and Partridge would then compare the average life spans of the females in the two groups. The first group would reveal the effects of both courting and mating; the second group would expose the effects of courtship alone.

Subtracting the effects of courtship from courtship and mating would leave the effect due to mating.

As simple as it sounds, there was one obvious problem to overcome. Fowler and Partridge had to find males that could court normally but were unable to mate. At first, they looked to see if there were any suitable sex mutants on the market. The *fruitless* mutant seemed a possible contender, a fly keen on courtship but reluctant to take things further. But *fruitless* was not ideal, because it will enthusiastically court males as well as females. Put a bunch of *fruitless* males in a bottle together and, within no time, they will be queuing up head-to-tail in a conga line of courtship.

There were plenty of other mutants to chose from. There were the children's favorites, *Ken and Barbie*, for example: mutant flies born with a bland bit of cuticle where the genitals should be. Then there was the Catholic mutant, *coitus interruptus,* a fly that gets cold feet about halfway through copulation. And most infamous of all, there was *stuck*. Penis problems mean that poor *stuck* can insert but cannot withdraw. Without assistance, *stuck* will remain glued to a female until he dies of starvation.

None of these mutants was exactly what Fowler and Partridge were looking for. All had idiosyncrasies that made them less than ideal. The absence of suitable mutants left only one alternative: to create flies that could court but not mate, normal males would have to be castrated.

It was time to bring out the castrometer.

In principle, the castrometer differs little from the transformer boxes used by model-railway enthusiasts. But instead of applying the electrodes to either side of the railway track, they are gently pressed against the nether regions of the male fly. The penis completes the electrical circuit and suffers melt-down as a consequence.

The device is a precision instrument, requiring a steady hand to achieve the best results. If you see smoke, you know you've pressed too hard. Not only will you have singed the penis, but most of the internal organs as well. Press too lightly, however, and the penis may still have some use left in it. The trick is to apply just enough pressure to melt and seal the genital opening, leaving behind a flat scar where a penis once stood.

The castrometer may seem like a return to the dark, Dickensian days of animal physiology, but, as cruel as it sounds, the flies received the best medical treatment available. All were given a general anesthetic before the surgery and most recovered their joie de vivre once the ordeal was over. Even their singing maintained the same even pitch. Apart from a little soreness between the legs, perhaps, the flies seemed perfectly fine, and within minutes of coming to, their interest in the opposite sex had returned to preoperative proportions. The flies could do everything normal flies could do, except mate.

Fowler and Partridge now had all they needed to distinguish the effects of courtship and mating on female life span. One group of flies were let loose with normal males, another group with the castrated males. And the results were striking. Females courted and mated by normal males lived shorter lives than females courted by the castrated males. In other words, there was something specifically about mating that was sending females to an early grave.

Exactly what part the seminal fluid played in this sinister story of sex and death was finally resolved in 1995, when Tracey Chapman and her colleagues confirmed the long-held suspicion that fruit fly semen was indeed a toxic time bomb.

The guilty verdict was made possible through a bit of very modern molecular biology. Tracey used a strain of male flies that had been genetically engineered to have defective accessory glands. The engineered flies were unable to make seminal proteins, but their production of sperm and other seminal chemicals remained unaffected.

Removing the proteins from the seminal brew prolonged female life span. Females who were lucky enough to mate with the engineered males lived about 50 percent longer than females who mated with the normal males. Here, then, was the final proof that seminal proteins could not only control females, but kill them as well.

It seems a perverse contradiction that semen—the substance that can give life—can also take it away. There is something not entirely logical about this state of affairs. It just doesn't make sense to kill your mate, at least not until she has finished laying eggs you have just fertilized.

Tracey thinks it unlikely that seminal proteins have evolved with killing in mind. Far more plausible is that toxicity is an evolutionary by-product, a side effect of chemicals whose primary purpose is to fight on the front line in the struggle for parenthood. Deliberate or not, the toxicity is real enough for females to mount an appropriate response of their own. The molecular war might seem one-sided, but females are far from sitting ducks.

Compared with the males, very little is known about the female side of the conflict. This bias in sexual knowledge is more a reflection of practical realities than sexism in scientific study. With males, seminal fluid is self-evidently the vehicle for their molecular offensive. But females are more tricky. Their defenses could be anywhere inside the body, and nobody is really sure where to start looking.

The details may be sketchy, but there is still plenty of circumstantial evidence in favor of a female defense. For example, not all of the male's accessory gland proteins seem designed for outright attack. Accessory gland protein Acp76A bears all the hallmarks of a molecular chaperone, hinting that it may escort and protect other seminal proteins as they make their way through enemy territory. What is Acp76A protecting the other proteins from, if not a female counterattack?

But the best evidence for female defense comes from experiments by fruit fly supremo Bill Rice, of the University of California at Santa Cruz. Not only has Rice demonstrated that females can deal with the male's molecular maneuvers, he has also shown that weapons on both sides of the sexual divide are constantly refined. Males and females, it seems, are locked in an evolutionary arms race.

Rice spent most of the late 1990s watching what happened to flies when he altered the normal evolutionary rules of engagement. In one experiment, he used special chromosomes called balancers, to create a strain of female flies that were unable to evolve. Rice then set up two groups of flies. In the first group, he paired up normal males with the evolutionarily hamstrung females. In the second group—the control group—he set up a conventional evolutionary engagement, with normal males and normal females.

When Rice came back to look at his flies after forty generations, he noticed some remarkable changes. With female evolution held in check, males seemed to have wrested control of the sexual conflict. Compared with the males in the control group, they had become much better at guaranteeing paternity, and were fathering more offspring. The toxicity of the seminal fluid had also increased. By the end of the experiment, the average life span of hamstrung females was about half that of the females in the control group.

Over forty generations, evolution had refined the male molecular arsenal, making seminal proteins more effective against both male competitors and females. Deprived of their evolutionary potential, females were unable to respond. What had been a war between two evenly matched powers had become a one-sided affair. It was as if the males were developing their latest Star Wars program while the females were still firing Scuds.

Rice's experiments show that males and females must play the same evolutionary games of cat-and-mouse as a parasite and its host. As soon as a male evolves a new line of molecular attack, the female must respond with a counterstrategy of her own.

Nobody is too sure how this evolutionary game translates into the cut and thrust of molecular combat. But one way of looking at the conflict is to think of it as a kind of chemical wrestling match for control of the female's behavior and physiology. It's not an unreasonable assumption to make, because many of the seminal proteins are thought to be mimics of the female's own hormones.

Hormones elicit metabolic changes in cells and tissues. The hormone molecules themselves do not get involved in any of the dirty work. They are merely chemical messengers delivering information to "receptors" on the cell surface. These receptors are molecular docking bays that enable hormones and other molecules outside the cell to communicate with molecules on the inside.

Seminal proteins will only excite a response in the female if they can dock effectively with the appropriate receptors on the surface of the female's cells. In other words, the male proteins are like molecular keys, whose effectiveness depends on a snug fit in the female's molecular locks. The evolutionary arms race, then, becomes a struggle between locks and keys. As fast as females can change the locks, males evolve new and more efficient keys to open them.

The chemical combat probably takes place throughout the female's body, but her genital tract and brain are thought to be the two principal theaters of war. One area where the fighting is thought to be particularly fierce is in and around the female's three sperm storage organs. This area has huge strategic importance because it is here that sperm is stored until getting the green light for the final leg of its journey to fertilize the female's eggs. Much of the molecular conflict is about wresting control of this light switch.

One of the accessory gland proteins known to operate in this area goes by the code name Acp36DE. This protein "corrals" sperm into the female's sperm storage organs. Males that lack the protein do not have their sperm stored very effectively, and their sperm loses out when it is in competition with the sperm of other males.

Another accessory gland protein, Acp62F, is thought to work the same sector. This mischief-maker may turn out to be the molecular equivalent of a loo chain, relaxing the muscles of the female's sperm storage organ and causing the sperm of rival males to be flushed away. Intriguingly, Acp62F bears an uncanny resemblance to a protein used by the Brazilian armed spider, *Phoneutria nigriventer*, to paralyze its prey. The leap from paralysis to poison is not a big one, and Acp62F has already been declared the chief suspect in the search for the seminal source of toxicity.

My own search for the source of toxicity takes me to a local pub. Tracey joins me for a pint and we chat briefly about her future research plans. She hopes to learn more about the strategies that female flies use to counteract the male molecular offensive. But she also has plans that go beyond the academic environment. Seminal proteins, she believes, may one day find a niche in pest control. Male insects could be genetically engineered to produce supercharged semen. Released into the wild, they would leave a trail of dead and frigid females in their wake. The Green lobby, I'm sure, would love it.

Before we have time to discuss the full implications of this idea, Tracey has to return to the laboratory to measure some more flies. I order another pint and begin to reflect on what a day in the laboratory has taught me, not only about sex but about the whole culture of fruit fly research.

Sexual conflicts in fruit flies certainly help to put our own petty squabbles into perspective. Never again will I argue about who's doing the washing-up or the position of the toilet seat. Then again, if fruit flies are any guide, perhaps our own verbal conflicts are nothing more than the tip of a very large iceberg? Maybe the really serious arguments are going on elsewhere? Until we learn a bit more about human seminal proteins, we cannot discount chemical warfare in our own species.

So the next time you are lounging in orgasmic aftershock,

with your postcoital cigarette, just consider what might be brewing up down below, as a battalion of sperm and their seminal support head north in search of eggs. It certainly gives the message of safe sex a new perspective.

Looking around the bar, I spot a couple of people I met earlier in the day, sitting in a corner, drinking themselves into oblivion. Also, there are a few old research colleagues, tropical biologists just flown in from Trinidad, Peru, and other exotic places. With their expensive and stylishly weathered mountain boots, and their feigned air of worldly wisdom, they look every inch the quintessential European explorers.

Their suntanned complexions stand in stark contrast to the sallow, pasty faces that I witnessed in the fly laboratory. There I perceived a slight sense of gloom, a solemnity perhaps born of the knowledge that however much work is achieved today, there will always be more tomorrow. The fly laboratory, with its temperature-regulated rooms controlling fruit fly production, its industrial-sized kitchen, and its lines of dutiful workers, is a factory in all but name. Beneath the gloss of polished white surfaces and expensive gadgets runs the spirit of the industrial cotton mill.

On my journey home, I pass the building where I have spent most of my day. High up on the fifth floor, the lights still shine brightly, a beacon of dedication in the darkness.

6

HELPING THE AGED

There can be few things in life more poignant than the sight of an elderly fruit fly trying to live up to the sexual reputation of his youth. His desire is still there, of course. That never goes away. But his body, ravaged by the unavoidable realities of old age, is no longer up to the job. His brash and arrogant swagger has all but disappeared. He prefers walking to flying. And his penis, never the most imposing of items at the best of times, has shrunk to a fraction of its former size. As hard as he tries, age has rendered him hopelessly ill-equipped for the rodeo ride that is fruit fly courtship.

The indignities of old age are familiar to us all. Brittle bones, graying hair, saggy skin, a fondness for bingo halls, knitted cardigans, and endless sitcom reruns—these are just some of the things we have to look forward to as we head inexorably toward our twilight years.

Faced with such a terrifying prospect, it is little wonder that antidotes to aging have always been in such great demand. The search for the secret of eternal youth is as old as history. But while there has never been any shortage of purported remedies, none has so far left a lasting impression. For a hunt lasting over five thousand years, ground-up monkey testicles, a diet of fresh air, and a Jane Fonda fitness video represent a relatively poor return.

Of course, there have been some things to cheer about. Twentieth-century improvements in sanitation and medicine mean that more of us are living longer than ever before. Traditional killers such as dysentery, cholera, tuberculosis, and diphtheria have largely disappeared from the developed world. In Britain, life expectancy has increased to about seventy-six years—a gain of thirty years since the Victorian era. Nevertheless, the idea that we may be able to slow down the onset of aging or treat its physical symptoms, and so extend the natural limits of the human life span, still seems hopelessly optimistic.

For humans, eternal youth may have to wait a while. In the meantime, however, we might do well to keep a close eye on the fly. Because while aging in humans seems as certain as ever, it's a different story altogether in fruit flies. Today, laboratory flies are living longer lives and growing old more gracefully than ever before. Soon the image of the elderly, arthritic fly may be nothing more than a poignant piece of nostalgia. Who knows, the fly may yet provide an escape route from our own mortality. Where there is hope, there is hype.

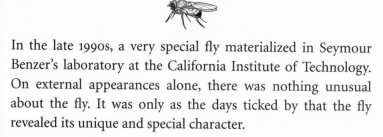

In the late 1990s, a very special fly materialized in Seymour Benzer's laboratory at the California Institute of Technology. On external appearances alone, there was nothing unusual about the fly. It was only as the days ticked by that the fly revealed its unique and special character.

On average, flies can expect to live for about fifty to sixty days in the laboratory. But at sixty days, this mutant fly was showing no signs of physical deterioration. With a youthful spring in its step, it was behaving like a fly half its age. Even after a hundred days it was still going strong. Eventually, age did get the better of the fly, but not before it had long outlived its laboratory peers.

The source of this fly's unique character was to be found inside the body. What distinguished the long-lived fly from the rest was a mutation in a single gene. The gene was named *methuselah* after the biblical character, who reputedly lived for 969 years. The fruit fly equivalent couldn't quite match that kind of longevity, but a 35 percent increase in average life span was not a bad return for a change in only one gene.

Longer life was not the only benefit of being a *methuselah* fly. The mutant also had a much stronger constitution than normal. When presented with a variety of stressful situations, the mutants fared much better than normal flies. Deprived of food, for instance, normal male flies could last for about fifty hours. But *methuselah* males could stay alive for eighty hours or more, an increase of more than 50 percent.

The mutant flies were also more tolerant of excessive heat. Higher temperatures increase the rate at which molecules inside cells vibrate. Turn up the heat too much and molecules may flip out of shape, or even break apart. Once protein molecules start to disintegrate, so does the fly. When normal flies were kept at 97 degrees Fahrenheit, they survived for about twelve hours. At the same temperature, the *methuselah* mutants managed to keep going for an impressive eighteen hours.

But the resilience of the mutants was most obvious when flies were exposed to the herbicide paraquat. Paraquat kills things because it generates highly reactive atoms and molecules called free radicals inside cells. These free radicals are chemical vandals that can quickly wear away the molecular fabric of a cell, and of an organism.

Twelve hours after being fed paraquat, normal flies were looking distinctly queasy and their movements were slow and lethargic. Hardly surprising, considering that their bodies were eroding from the inside. After forty-eight hours, nearly 90 percent of the normal flies were dead. But it was a completely different story in the *methuselah* camp; twenty-four hours after their paraquat supper, the flies were as perky as spring lambs. Things did begin to go downhill for the flies soon after, but after forty-eight hours, more than 50 percent of the flies were still alive.

The *methuselah* fly offered a revealing insight into the nature of aging and longevity. Severe stresses in the form of starvation, excessive heat, and exposure to herbicide inflict biochemical damage on the fruit fly body. The *methuselah* mutant suggested that slower rates of aging and a longer life come to those who are good at resisting or repairing this kind of damage.

The *methuselah* mutant is just one in a long line of discoveries that have seen the fly propelled to the forefront of aging research. While the fly has yet to solve the riddle of aging, it has, at least, brought much-needed empirical clarity to some extremely choppy theoretical waters. Over three hundred different theories of aging have been proposed over the years, and the biology of aging remains littered with discarded and sometimes conflicting ideas. But one thing now seems clear: it is extremely unlikely that aging has a single cause. On the evidence gleaned from the fly, there are probably many interacting factors that contribute to our decline.

Before it moved indoors, the fly had little chance to experience the pleasures and the pains of an aging existence. Out in the wild, the daily risk of death is so high that few flies get the chance to grow old. Predators, parasites, viruses, fungi, and disease-causing bacteria are always queuing up to get a taste of the fruit fly body.

The laboratory offered sanctuary from all this madness. Free food, a warm bed for the night, and no predators or parasites gave the fly the chance to follow its sex drive without distractions and live life to the full. In fact, those who argue that captivity is a cruel life for an animal would do well to consult the fly. A wild fly might consider itself lucky if it were still alive ten days after its birth. In contrast, laboratory living could give the fly dozens of extra days on Earth.

The length of a fruit fly's life depends not only on the individual, but also on the temperature. Fruit flies are "cold-blooded"; in other words, their body temperature fluctuates with the temperature of the environment. In practical terms, this means that the fly's pace of life is at the whim of the environmental temperature. When it gets hot, their metabolism speeds up and the flies become hyperactive. When the temperature drops, their metabolism slows down and they become dopey.

In 1917, Jacques Loeb, Thomas Hunt Morgan's friend and onetime colleague at Bryn Mawr College, found that higher temperatures equated with shorter fruit fly lives. Flies reared at 68 degrees Fahrenheit lived for an average of fifty-four days. At 77 degrees Fahrenheit, the average life span was cut to thirty-nine days, and at 86 degrees Fahrenheit, it dropped still further to a miserly twenty-one days.

The observation that higher temperatures—and higher rates of metabolism—equated with shorter life spans led to the development of the "rate of living" theory of aging. Life was like a song, with a finite number of beats. Different life spans were the result of different interpretations of the song. While long-lived animals went for the slow ballad, others, like the fly, preferred a faster, techno rhythm.

The theory proved popular. Its prediction that the relative metabolic rates of different species should reflect their relative life spans, seemed to fit with the data. Fruit flies, rock stars, even warm-blooded animals, were neatly accommodated by the theory. Small mammals such as mice and shrews, for example, had much higher metabolic rates and shorter lives than larger mammals, such as humans and elephants.

But the theory wasn't perfect. Metabolic rate is not always a reliable indicator of life span, particularly if you happen to be a bird. A pigeon, for example, has a much higher metabolic rate than a rat, a mammal of comparable size, but a pigeon lives for about thirty years to the rat's four.

While the rate of living theory has fallen out of favor, a link between aging and metabolism has not disappeared altogether. More recent ideas on aging have focused on metabolic by-products, rather than the rate of metabolism per se. These days, hazardous by-products of oxygen metabolism are some of the chief suspects in aging.

We need oxygen to stay alive. This oxygen, of course, comes from the atmosphere. Within seconds of breathing air into our lungs, oxygen has been distributed, via the bloodstream, to the billions of cells that make up the body. Just as oxygen is needed to burn fuel and produce heat, the cells of our bodies need oxygen to "burn" the food we eat, to produce energy.

Inside cells, this combustion takes place within specialized compartments called mitochondria. Energy is produced by way of an extremely long-winded chain of chemical reactions on the inner surface of the mitochondria. Oxygen's job in all this is to mop up the free electrons that shoot off the end of the chain. In the interests of health and safety, it is important that each oxygen molecule picks up four electrons. If it picks up only one electron, the oxygen turns into a highly unstable molecular thug—a free radical. In this state, the oxygen has the potential to cause enormous damage. Free radicals are always looking to provoke a reaction from any molecule they happen to bump into. They have only one thing on their minds—electrons—and will stop at nothing until their electron lust is satisfied.

When free radical formation spirals out of control, then you've got a fire on your hands. Clearly, fire is something that a cell would rather avoid. The combustion that goes on in cells is a much more sedate and controlled affair than the kind that goes on in your hearth. Cells make every effort to prevent free radicals from forming in the first place. Inside the mitochondria, an enzyme holds the oxygen in a molecular straitjacket and only lets go when the oxygen has picked up four electrons. That, at least, is the theory. In practice, 2 to 3 percent of the oxygen always manages to struggle free.

Let loose, free radicals rampage around the cell like molecular hooligans. Proteins, DNA, and lipids are just some of the molecules that can have their chemical and physical faces rearranged by the unwanted attention of a free radical. The DNA alone can expect to receive about ten thousand molecular punches each day.

Fortunately, cells are not defenseless against free radicals.

They carry a whole suite of enzymes whose function is to clean up and repair any molecular cuts and bruises, or put the free radicals out of action before they get a chance to cause damage in the first place. For example, antioxidant enzymes, such as superoxide dismutase and catalase, form a molecular police force against the free radicals. Their job is to intercept and neutralize the free radicals before they start any trouble.

Many antioxidants are produced "in-house," but they can also be consumed as part of a balanced diet. Vitamins C and E, for instance, are well-known antioxidants. Vitamin E, found in nuts and green vegetables, specifically protects the molecules that make up a cell's delicate membranes from free radical damage.

Just how valuable antioxidant enzymes can be to an organism was shown in the early 1990s, when fruit flies were genetically engineered with extra copies of the genes for catalase and superoxide dismutase. The additional genes not only boosted the physical fitness of the flies, they also increased their life span. The genetically engineered flies lived about 30 percent longer than a control group of flies that had not received extra copies of the antioxidant genes. What is more, the proteins of the genetically engineered flies showed far fewer signs of wear and tear than those from the control group. The experiment provided compelling evidence that aging can be blamed, at least in part, on free radicals. But it also showed that antioxidant enzymes can provide a powerful antidote to the physical symptoms of growing old.

With the fly painting such a rosy picture of antioxidant enzymes, it is little wonder that antioxidants have become a desirable item in the diets of the health-conscious. But although dietary supplements of antioxidants have been shown to increase the life span of fruit flies, in humans their effects are more ambiguous. A sprinkle of antioxidant on your bran flakes does not seem to guarantee a longer life, although the reasons are still unclear. Perhaps increasing your intake of only one or a few enzymes upsets the balance of a more complex and interactive enzyme system. Or maybe the antioxidants are broken down in the gut before they have a chance to get inside cells and do their work. Whatever the explanation, there seems to be no quick fix on offer, just yet.

Oxygen is not the only source of free radical formation inside cells. Ultraviolet light and other forms of radiation, the toxins in tobacco smoke, dozens of environmental pollutants, herbicides, and heat can all prompt the production of these molecular troublemakers. Free radicals are part and parcel of life. Any one cell will produce millions of them every day.

Evidently, the fact that we do grow old suggests that the suite of enzymes that make up a cell's defense system are good, but they are not perfect. Free radicals can hit home and make their presence felt, and damage to DNA can go unnoticed, or be repaired incorrectly.

On its own, the odd bit of molecular vandalism here and there is unlikely to pose a serious threat to the well-being of the cell. But add up all the little acts of violence over days, months, and years, and they develop into a far more serious concern. The molecular bricks and mortar of a cell will begin to weather and erode. The channels of communication that exist within and between cells will begin to break down; and the corrosion of biochemical machinery will send energy production into gradual decline.

If that isn't gloomy enough, then there is always cancer to look forward to. The gradual accumulation of free radical damage provides an explanation of why diseases such as cancer are just one of the many calling cards of old age. Cancers are triggered by mutations in genes that regulate the growth and division of cells. The longer life goes on, the greater the chances of one of these genes being hit and damaged. When these genes mutate, cells can lose all sense of self-control and proliferate to form tumors.

Repair enzymes are there to protect against these kinds of mishaps. For example, DNA "proofreading" enzymes move along the double helix, detecting and correcting chemical defects in the DNA. But the genes responsible for the cell's defense enzymes are themselves susceptible to damage. Once the cell's defense mechanism has been disrupted, then there is little to stop the molecular violence spiraling out of control.

Cancer is not the only disease of old age in which free radicals may play a starring role. Neurodegenerative conditions such as Parkinson's disease and Alzheimer's disease have also been linked to free radical damage. The cells that make up the brain and nervous system are especially vulnerable because they metabolize much more oxygen than the average cell.

In fact, although all cells are susceptible to free radical damage, genetic engineering experiments with the fly suggest that it is free radical damage in the cells of the nervous system that have the biggest impact on life span. Flies engineered with an extra copy of an antioxidant gene that is switched on only in the brain and nervous system live as long as flies that have an extra gene switched on throughout the body.

Nerve cells may well lie at the heart of aging when it comes to the effects of free radical damage. But there is more to a long life than quashing the violent tendencies of free radicals. If you are looking for an extended time on Earth, then you might also want to consider what heat can do to your cells.

Heat can inflict injury to cells by increasing the production of free radicals. But heat can also damage molecules directly. Fortunately, cells are not defenseless against the occasional hot spell. They can rely on an army of heat-shock proteins— so called because they are activated by high temperatures—to patch up and replace molecules that have become distorted and misshapen by heat.

To illustrate just how important these proteins can be to a longer life, take two groups of flies. Genetically engineer one group with extra copies of a gene for a heat-shock protein. Leave the other group as they are. Stick both groups somewhere hot and sticky and monitor them each day. You should observe that while the genetically modified flies continue to party and soak up the sun, the nonmodified control group are soon dropping like, er, flies. The message seems simple. If you can't stand the heat, get yourself some more heat-shock proteins.

Evidently, there is more than one route to a longer life. You could boost your antioxidants, perhaps, or add an extra dose of heat-shock protein. And it would always be wise to keep an eye on the state of your DNA repair enzymes. But why not go the whole hog, and boost your entire defense strategy? That, after all, is what the *methuselah* fly seems to have done.

The *methuselah* mutant has a heightened resistance to a wide variety of environmental stresses. A change in a single gene has enabled it to maintain a stiff upper lip in the face of heat, starvation, and free-radical-forming herbicides. Nobody knows for sure how the protein product of this gene weaves its mortal magic. The protein seems to sit inside the membrane of cells, from where it probably oversees and directs the overall defense strategy. Its effectiveness comes, perhaps, from the speed or efficiency with which it coordinates a molecular response to any kind of stressful situation.

The *methuselah* mutant appears to present an interesting biological paradox. If a mutation produces a new version of a protein that makes a fly better able to resist stress and live a longer life, then why has this mutant not been "discovered" by evolution already? If it confers such a clear advantage, then why is the *methuselah* mutant not already widespread in natural populations of the fly?

There are two answers to this puzzle. First, it's possible that a *methuselah* mutant has never cropped up in a wild population of flies. And, second, even if a mutant has appeared, it would not necessarily have an advantage in the struggle for existence. You only have to look at the huge variation in the average life span of different species of plants and animals to see that a long life per se is not universally advantageous. What may be desirable for humans does not necessarily make good evolutionary sense for flies.

But if aging is simply the accumulation of molecular damage, then why do different species of animals and plants have such different life spans in the first place? Why is it that we can live for seventy-five years or more, while the fly is lucky to last a few weeks? After all, we breathe the same air.

Although things might seem equal from the outside, look

inside the body and you begin to see some alarming dispari-
ties. All animals and plants have repair mechanisms that can
patch up molecular damage inside cells. But the standard of
service on offer varies enormously among species. Humans,
for example, have a very efficient repair system, one that can
spot most of the problems that arise. The system in mice is less
efficient. Nevertheless, it's good enough to keep them on the
Earth for a year or two. Fruit flies, however, seem to fare less
well. If you think of a repair mechanism as a tool kit, then evolution seems to have given the fly a plastic screwdriver from a
box of Cracker Jacks. Although they have their uses, they are
clearly not designed for the long term.

But variation in the competence of repair services is only a
partial answer to the question of why different species have
different life spans; it still leaves us with the question of why
flies evolved a low-grade service in the first place. Why did they
get such poor quality, while we got the gold star treatment?

One popular idea is that differences in repair services and
life spans may reflect different evolutionary solutions to the
problem of a finite energy budget. All organisms are limited in
the total amount of energy they can generate. Some of this
energy must be devoted to reproduction, the rest to basic body
maintenance. How the energy budget of a species is divided up
seems to depend on its risk of accidental death—the chance of
being eaten by a predator, succumbing to disease, being
squashed underfoot, walking under a bus, falling off a cliff, or
suffering any other kind of terminal mishap.

If, for example, you are the kind of species that routinely
disappears down the throats of hungry predators, then it
would make little evolutionary sense to invest too heavily in
body maintenance. Few bodies would be around long enough

to reap the benefits. Far better in these circumstances to do what the fly has done: devote most of your energy budget to reproduction, and start bonking as soon as you are born. The downside is that you have little energy left over for basic maintenance, and age very quickly as a consequence.

In contrast, if you are an animal with little to fear from the outside world, such as a human being, an elephant, or a tortoise, then rapid aging no longer makes sense. In this situation, it would be better, from an evolutionary standpoint, to invest more in basic maintenance. You will reproduce more slowly, but live longer in the bargain.

But if you want to live a really long life, then it seems that you have to abandon sex altogether. Asexual species such as sea anemones show no visible signs of aging at all, and seem able to go on living indefinitely. Likewise, historical records indicate that many kinds of trees have the potential for immortality.

In one sense, we too are immortal. After all, our sperm and eggs perpetuate the family line. The problem is that the bits of us that we're most attached to—our bodies—are not. Obviously, bodies are immensely valuable to us as individuals, but in an evolutionary sense, the body of a sexual organism is merely a caretaker for the sex cells—the sperm and eggs. The body's job is to keep an individual alive long enough to reproduce. After that, body cells can accumulate mutations and age because they have fulfilled the job they were supposed to do. Let's put it another way: once evolution made us sexual, it also made our bodies disposable.

In asexual species, there is no distinction between sex cells and body cells. All the cells of an asexual organism are effectively equivalent. Reproduction is an extremely dull affair in which a single cell, or group of cells, splits off from the main body of the organism to form clones—identical genetic copies—of the "parent." Dull it may be, but with no separate "body" to dispose of, the upside is a long and potentially eternal life.

If an asexual life is the key to an ageless existence, then perhaps we are not as far away from immortality as we might think. When I say "we," I am, of course, referring only to the female half of the population. To all intents and purposes, males are irrelevant to the asexual equation.

Dolly the sheep has taught us that females could probably go asexual tomorrow. All it would require is a bit of technological tweaking and a more sympathetic government. You could take a cell from some part of your body, whip out its DNA, stick the DNA inside one of your empty egg cells (one that has had its own DNA removed), and put the egg back into your womb. Nine months later, all being well, a baby clone of yourself would emerge, kicking and screaming, into the world.

But as attractive as the idea may seem, it's unlikely that this Utopian scheme would have much of an impact on aging. Even without males, females would still be biologically sexual organisms. To evolve true asexuality, in the sea anemone sense, females would have to do away with eggs entirely. They would also have to do away with genitals, breasts, lips and hips, and all the other features that are superfluous to an asexual lifestyle. These sorts of evolutionary changes are not going to happen overnight. It might take thousands of generations of selective breeding to turn a sexual population into an asexual one. Even so, the enticing prospect of pulling off an arm and watching it grow into an identical copy of yourself might justify the wait.

Going clonal is, perhaps, a step too far for most people's tastes. After all, would you be prepared to give up sex for an immortal life? We're not just talking here about giving up mixed-sex interactions. Selective breeding for asexuality means that all erogenous zones would freeze over like the polar ice caps. Life would indeed be long. It would also be extremely boring.

So rather than doing away with sexual reproduction altogether, how about just delaying it a bit? Selective breeding experiments with fruit flies have shown that if you delay the age at which flies reproduce, then, within a few generations, flies are living longer and healthier lives.

By selecting old flies as the parental basis for the next generation, you are, in effect, selecting for the ability to remain fertile into old age. Put another way, selecting for delayed reproduction is equivalent to selecting for slower aging. It certainly works in flies, and it might even work for humans. In women, there is evidence that old age goes hand in hand with a delayed menopause. Women who live to be a hundred remain fertile for longer than average.

Longer life in selectively bred flies brings other benefits. Not only are the flies better at resisting a variety of environmental stresses, they are also more athletic. When it comes to sustained walking and flying, they can easily outpace and outlast normal flies of the same age.

But these benefits do not come free of charge. The flip side of a longer and healthier life is a lower fertility in early life. The longer-lived flies seem to evolve a new energy budget, investing more in body maintenance at the expense of reproduction. Again, there are hints that these types of effects may exist in humans. Historical records from the English aristocracy indicate that women with greater longevity had lower overall fertility.

The benefits of delayed reproduction can be thought of not just in terms of energy budgets, but also in terms of genes, genetic profiles, and changing patterns of natural selection. Genes persist in populations because they are good at being transmitted from one generation to the next. They do not persist so people can live happily into old age. Once an individual has reproduced, its genes no longer have much interest in what happens either to them or their owner.

Natural selection ensures that genes that reduce an individual's chances of survival are less likely to persist in a population. But the intensity of natural selection will depend on the age at which the gene manifests its effects. In prereproductive age, for example, natural selection is fierce at weeding out defective genes. But once an individual has passed the point of reproduction, selection is more or less a spent force. So any genes that have deleterious effects in the postreproductive years will persist in a population.

This reproductive distinction can be illustrated with two human genetic disorders, progeria and Huntington's disease. Both are devastating diseases that kill. Progeria causes premature aging during childhood, and death normally occurs in the teens. Huntington's disease, a neurodegenerative disorder, does not manifest its effects until well into middle age.

Progeria is extremely rare because individuals who carry the gene normally die before they have a chance to pass it on to the next generation. In contrast, Huntington's disease is relatively common. By the time people contract the disease, they are usually past their reproductive age: the gene has already found its way into the next generation.

The gene for Huntington's disease represents an extreme example of a late-acting deleterious gene. But there are probably many genes present in human populations that have much milder effects. They may even be beneficial early in life but turn nasty later on, contributing to the physical deterioration associated with aging.

Whether genes have severe or mildly deleterious effects, the principle is the same. Provided that these genes manifest their effects after reproduction, they will not be weeded out by natural selection. But by delaying the age of reproduction, you extend the period of intense natural selection. All these annoying deleterious genes, for so long concealed, are now exposed to the ruthless hand of selection. Eventually, over many generations, they will be weeded out. The result is longer and healthier lives.

If delayed reproduction can benefit the lives of fruit flies, then why not humans? There is no reason, in theory, why humans would not reap the rewards of delayed reproduction. But remember, this is a long-term evolutionary project. Delayed reproduction has no immediate effect in an individual's lifetime. It would be a massive experiment in social and genetic engineering, lasting many generations and demanding the cooperation and participation of entire populations. Each generation, procreation would be put exclusively in the hands of the old but fertile. Anyone in their early twenties wanting to get married, settle down, and have a family would be banned from doing so. Can you imagine how difficult it would be to police and enforce? And all for the benefit, not of the present, incumbent generation, but of generations that lie way ahead in the future.

So perhaps the best solution for a longer life is to forget about sexual interactions altogether. As the fly has shown, celibacy can work wonders for your life span. Flies that are denied the opportunity to engage with the opposite sex live longer lives than those reared in mixed-sex neighborhoods. Sadly, however, there is little evidence that the same drastic measures would work in humans. Men and women who remain childless throughout life do not live longer lives. And neither do monks and nuns, who embrace a life of celibacy.

With or without sex, there are plenty of other avenues of interest. Instead of abstaining from sex, we might be better off cutting down on how much food we eat. After all, in flies and many other animals, a restricted diet can increase life expectancy by a third.

Dietary restriction does not mean malnutrition. Restricted diets contain all the right vitamins and minerals; only their energy value is lower than normal. Energy restriction certainly seems to have made a difference to the lives of the people on the Japanese island of Okinawa. The Okinawans consume about 20 percent fewer calories than people living on the Japanese mainland. With 185 people in every million reaching the age of a hundred or more, Okinawa has the highest proportion of centenarians in the world.

Nobody is quite sure how dietary restriction extends life span. Nobody is even sure whether long life on Okinawa is the result of reduced energy intake or some other peculiarity of the diet. The Okinawans, like all Japanese, have a diet that is rich in fish oils, soy products, and vegetables—foodstuffs that are known to have beneficial effects on health. Perhaps this is the reason why the Japanese have a life expectancy of eighty years—the highest of any population in the world.

The study of aging is still in its first flush of youth, and human aging remains something of a mystery. But at least the fly has helped to identify areas of biology that deserve a second look. Of course, the fly is not alone in its role of scientific prophet; there are dozens of pretenders to the Methuselah throne. Rats, mice, monkeys, and nematode worms, to name but a few, are all enduring dull diets, enforced celibacy, and a host of other hardships in the human pursuit of a longer life.

Even if a cure is still some distance away, the glut of new aging theories suggests that the search for the secret of eternal youth is more fervent than ever. Personally, however, I can't think of anything more horrific than a cure for aging. Can you imagine it? All those endless repeats on television—only more of them.

Growing old has always been an inevitable fact of life. The question is, would you want it any other way?

7

HOT SPOT ON HAWAII

The jumbo jet made its final, cumbersome maneuvers for landing. Again and again, it rolled first one way and then the other, constantly correcting its course as it homed in on a pin-sized strip of tarmac three thousand meters below. Descending rapidly, the aircraft's engines swallowed great gulps of sky, reducing thousands of insects, riding high on freak upwellings of air, to sloppy sprays of confetti.

A jolt of turbulence partially roused me from my slumber. For a moment I sat stupefied in my seat, while my mind explored the isthmus that connects the twin continents of sleep and non-sleep, the place where the marshmallow world of dreams blends with the metallic reality of full-blown consciousness.

Staring out of the window, my glazed eyes settled on the two engines suspended beneath the oscillating wing. I was instantly gripped by a sense of unease. The engines could have been two ripe oranges just waiting for a gust of wind to blow them from their tree. I felt sure they were going to fall off at any moment. Frantically, I began counting the rivets for reassurance.

Panic set in when I spotted a rivet hole without a rivet. Was this a deliberate modification, planned and discussed in an engineer's drawing office? Or was the missing rivet an oversight, an error, a testament to human fallibility? Even the people who weld rivets onto aircraft wings must occasionally make mistakes. A hangover, a lost love, a bereavement in the family, any one of these things could cause an ordinary mortal to overlook a rivet. At ground level, a missing rivet was perfectly understandable. But from where I was sitting, suspended in a sardine can high in the sky, forgiving was the last thing on my mind.

A courtesy breakfast bun brought me back down to earth. Thoughts of rivets disappeared as I devoted all my efforts toward chewing and swallowing the cementlike cake. When I next glanced out of the window, I felt a warm glow as the unique landscape of Hawaii came into perfect view. Brooding volcanic peaks towered over the islands, and their handiwork was plain to see. Much of the land surface was covered in a solid carpet of black pumice, but patches of lush tropical rain forest stood defiantly in the areas that the lava had left untouched. It was a mesmerizing, polka-dot landscape, where life and death sat side by side. Only when the aircraft sank lower in the sky did this mix of extremes dissolve away, as the skyscrapers of Honolulu heralded the arrival of a more familiar and prosaic ecology.

Hawaii is one of the world's special places. But not for the reasons you read about in the tourist brochures. Forget the fine yellow sand, the clear blue sea, and the world-famous surf. Forget grass skirts and the garlands of freshly cut flowers. Forget Elvis, the lurid shirts, and *Hawaii Five-O*. And forget the languorous music that sways to the rhythm of the breaking waves. Forget all this stuff. The real attraction is the wildlife that makes a living in Hawaii's schizophrenic landscape.

Hawaii's eight islands are home to at least 22,000 species of animals and plants, almost a half of which are found nowhere else on Earth. One thousand species of flowering plants, over ten thousand species of insects, and sixty species of birds are unique to the islands. Hawaii, like the Galápagos Islands to the south, is a Darwinian dreamland, a place where the origin of species has run riot.

Odd, then, that Hawaii can't match the eco-tourist cachet of its Pacific neighbor. Is this because Hawaii has nothing as photogenic as the Galápagos tortoises and marine iguanas, or anything as historically resonant as Darwin's finches? Or is it because Hawaii's most outstanding example of evolution in action is the fruit fly, an insect with a woefully poor track record in public relations?

About a half of all the world's fruit fly species live on this remote volcanic archipelago. That's about a thousand species, in an area not much bigger than your average English county (all of Britain, by contrast, has about thirty species). Worldwide, something like 0.0001 percent of all insect species are fruit flies. On Hawaii, that figure is more than 10 percent. When it comes to animals, the Galápagos may have the glitz and the glamour. But if you're more interested in the origin of species than in photo opportunities with an oversized and outdated reptile, then Hawaii is the only place to be.

The first fruit fly probably arrived in Hawaii about thirty to forty million years ago. Colonization could have come from a single pregnant female or a small band of migrant flies; nobody knows. Nor does anybody know where the fruit flies came from, or their mode of transport. But one thing's for sure: my trip to

Hawaii, like the fly's, was a once-in-a-lifetime experience. I was on a pilgrimage to the Mecca of evolutionary biology. So, too, were hundreds of others, flying in from all over the world to attend a conference on evolution. For a young and enthusiastic Ph.D. student, it was a dream come true.

What made the trip even more pleasurable was the fact that I was not paying for it. I had managed to convince my grant-awarding council that they should foot the bill for this extravagant adventure. As generous as it seemed, the council's financial support came with strings attached. I could go to the conference in Hawaii provided that I presented a talk there on my council-funded research. This didn't seem like a bad deal to me. Although I had never given a talk at a conference before, I had given plenty of seminars within my own university department.

My talk was scheduled for 8:40 A.M., a time when I'd normally be tucked up in bed. Slightly more worrying, however, was the location. The organizers had put me in the largest of the conference venues, a cavernous two-thousand-seat auditorium that dominated the university campus. But my slot was still days away. I put my worries to one side and immersed myself in the conference atmosphere, and the glut of talks on offer.

Conference talks are a bizarre academic ritual. People turn up, ostensibly to hear the latest scientific breakthroughs. But you look around the average audience and hardly anyone seems to be listening. Only a big-name speaker or a sexy scientific topic can guarantee more than half the audience's attention.

Look more carefully and you will find that levels of attention are not uniform throughout the audience. Most of the time, they are stratified, like layers of sedimentary rock. In the shadows at the back is the layer of least interest, populated by people reading newspapers or sleeping off hangovers from the previous evening.

Those seated toward the middle will have half an ear on the speaker, but they will probably be doing other things, too, like preparing their own talk. Only at the front will there be people giving the speaker their full attention. Here you will find the furious note-takers who hang on every word. The speaker only has to cough for their pens to swing wildly into action.

There are several reasons why most of the audience is invariably indifferent. First, and most obviously, is the fatigue factor. An academic talk works best as an event on its own. String it together with a whole load of other talks and you end up with a marathon of monotony and an overwhelming sense of ennui.

Then there is the presentation itself. Research scientists are not trained in the art of public speaking. They are often too nervous, too arrogant, or too dull to bridge the communication gap. For the audience, the struggle to understand can be a mentally draining experience. Sleep, therefore, is not just a lazy cop-out. It is frequently a medical necessity.

But perhaps the main reason for the apathy of your average conference audience is the fact that talks are only a sideshow to the main event. They are advertisements competing for audience attention in the free market of ideas, the hors d'oeuvre to the main course of frenzied networking that takes place afterward, the format that gives structure to five days of booze and banter.

Half an hour before my own talk, however, this thought was providing little in the way of comfort. Whether anyone in the audience was going to be listening or not, it made no difference to my state of mind. Nerves were building rapidly, like gathering clouds before a storm. It was the arrival of self-doubt, the first cousin of fear, and an early hint that my mind was becoming populated by sensations beyond my control.

Little did I realize that this uncomfortable state was merely the launch pad of an increasingly fraught mental journey. I saw the hand of the clock tick on a minute more and stepped forward, off the edge of the diving board, falling headlong, in a spiraling descent, into the black depths of anxiety. Nine-point-eight, nine-point-nine, nine-point-nine; the judges' marks said it all. It was a great dive.

The impact hit me like a drug. Suddenly, a familiar mental landscape gave way to a desolate and sinister version of existence. Old certainties grew arms and legs and four devilish heads that feasted on the flesh of my fears. Fear begat fear with the fecundity of a fruit fly. I struggled to find mental footholds, to halt my slide further into the abyss. But it was no use. A demented orchestra was blasting away in my head, on a suicide drive toward the white noise of blind, naked panic.

I looked around the auditorium, hoping to establish eye contact, to receive a smiling face, a connection with the solid certainties of the old world. But there was nothing to cling to, bar an overwhelming sense of alienation. I stared up at the ceiling, into the black and infinite distance, listened to the big booming echoes that emanated from the gates of hell, and finally glanced at the vast black stage, awaiting my execution. Maybe I was already dead.

Clearly, the trip had turned bad.

For a moment, things eased off a bit, allowing me a brief flash of clarity. There was an exit from this world. I could get up right now and walk straight out of the auditorium, along the path that led out of the university campus and down toward the sea. But then where? I was surrounded by thousands of kilometers of ocean. There really was no exit. It was all an illusion, a fiendish trick.

I was aware of the chairman announcing my name, his

words completely out of sync with his lips. There was a silence that probably lasted a second but felt like an hour. Acting on some primordial motor instinct, my legs took me to the foot of the stage. I stared at the chairman, who blessed me with a removed but benevolent smile, the kind given to a man about to meet his maker, and walked up the steps onto the stage. There was a lectern about a mile away. When I reached it, I clung on for support. With the clip-on microphone dangling crookedly from my shirt, I began.

Weeks before, when preparing my talk, I had planned to open with a joke. Given the current circumstances, I decided to shelve it; there was no confidence left in me to pull it off. Much to my amazement, however, words did come out of my mouth, and pretty much where I expected them. Within five minutes or so, I had relaxed sufficiently to begin making a few quips about the substandard quality of my projector slides. The talk went smoothly. I even dealt comfortably with the questions afterward. When it was all over, I sauntered off the stage, as someone else took their turn.

Later that day I reflected on the morning's events. All that fear, all that anxiety. All for a bloody talk about moths! It's a good thing I wasn't talking about something important. Like psychosis. But, despite the trauma, the talk had gone well. So well, in fact, that one member of the audience would later offer me a job. Some consolation, I suppose, for a traditional welcome into the world of panic attacks.

Standing on that Hawaiian stage all those years ago, waffling on about moths in South Wales, I can remember being hit by a huge sense of the absurd. There was nothing odd about the subject of the talk. It centered on a topic many would regard as the big question in evolution: speciation—the origin of species. No, the topic was fine. It was the context that was so incongruous. When it came to studies of speciation, moths in South Wales seemed trivial compared with fruit flies in Hawaii. Trying to impress on the audience the importance of my work was a bit like trying to play up the attractions of a beach holiday in Blackpool to a delegation of tourist reps from the Seychelles.

Remote islands and archipelagoes have always received a disproportionate amount of attention from evolutionary biologists, and it is not difficult to see why. Islands are great places for the origin of species, breeding grounds for diverse collections of unique forms. The Hawaiian fruit flies are one example, but you can take your pick from the lemurs of Madagascar, the moas of New Zealand, the finches of the Galápagos, and countless others. Islands, as evolutionary biologists never tire of saying, are natural laboratories for the study of speciation. It just so happens that many of them are also great places to get suntans.

At over 3,500 kilometers to the nearest mainland, Hawaii is the most isolated archipelago on Earth. The islands have always been remote. This is no broken-off piece of continent that went walkabout in the Pacific. The islands were delivered from the sea by volcanic eruptions on the ocean floor.

Being stuck out in the middle of the Pacific meant that Hawaii was never an easy place to get to. It must have been more or less potluck who made it across the vast expanse of unbroken ocean. The original colonists were probably a paltry hodgepodge of species from the Pacific rim. But once established, these few seeds were somehow able to grow and blossom into the species blooms we see today.

There are eight islands altogether, which form a chain stretching from Kauai, in the northwest, to the "Big Island" (also known, confusingly, as Hawaii) in the southeast. All the islands are fixed to the Earth's vast Pacific plate, which drifts northwestward, on the liquid mantle below, at a rate of about nine centimeters a year.

Successive Hawaiian islands have emerged from the ocean as the Pacific plate has moved over a fixed "hot spot" within the Earth's mantle. The hot spot is a place where upwellings of molten magma can melt the overlying plate, spew up into the sea, and solidify into islands of volcanic rock.

Kauai, being the most northwesterly of the islands, is also the oldest, first appearing about six million years ago. The Big Island, to the extreme southeast, is the youngest, at about half a million years old. In fact, its southeastern corner, an area of intense volcanic activity, sits right over the hot spot and is still under construction.

Although the age of the islands has never been in doubt, concerns were raised a few years ago when the DNA of Hawaiian fruit flies came under scrutiny. As expected, all the Hawaiian species were found to be more closely related to one another than to any species outside the islands. This was consistent with the idea that the Hawaiian species had evolved in situ. But the survey also uncovered an apparent contradiction. The genetic data indicated that the Hawaiian flies originated from a colonization event that took place at least *twenty-five* million years ago. And yet Kauai, the oldest island, is only about five or six million years old. In other words, none of the current islands existed when the fruit fly first arrived in the area. Did the fruit fly colonists float around the Pacific in makeshift life rafts for twenty million years, waiting for an island to grow out of the sea? Or did they form an aquatic race that later rose from a fruit fly Atlantis to colonize the land? Alas, neither seemed plausible.

The most likely explanation is that there were other islands, older than Kauai, that have already been and gone; islands that erosion has returned to the sea. Judging by the rate at which the Pacific plate moves, the island on which the first fruit flies landed is now believed to be buried beneath the waves, somewhere near the Midway Islands to the northwest, over three thousand kilometers from the current crop of Hawaiian islands.

So it looks as though the history of fruit flies in Hawaii has necessarily been a history of island-hopping. As old islands slipped below the surface, flies were forced to move on, to new and younger islands. These journeys were not to be taken lightly. Island-hopping in Hawaii is not like popping down to the corner shop for a pint of milk. Some of the islands are separated by considerable distances. The Big Island, for example, is a fifty-kilometer flight from Maui, its nearest and older neighbor. That's no picnic for a small insect like a fruit fly. Chances are that very few flies would complete the journey.

The current distribution of fruit fly species in Hawaii is consistent with infrequent migration between islands. Each island has species that are found on none of the other islands. What is more, the closest relatives of these endemic species consistently map to the preceding, older, island in the chain. It looks as if each island has been a stepping-stone, with bursts of new species appearing at every stage along the way.

The key to these new species blooms may lie in the small size of the island-hopping populations. Because they are few in number, migrants will not always carry a representative sample of a species' genes. The population they help to establish could, by chance, have a different genetic profile from the one they have just left. If the founding population also finds itself in a different sort of environment, natural selection could further accentuate genetic differences.

One way to illustrate these effects is to think of the original population as a book; a novel, for example. A founder event—the migration of a small group of individuals to a new island—is like ripping a few pages out of the book and then handing them to someone else, someone who has never seen the book before, and asking them to complete the story. The new version is unlikely to bear much resemblance to the original. Likewise, the new island population may take on a form different from the old. With little or no migration between islands to harmonize the diverging genetic profiles, the origin of a new population could turn into the origin of a new species.

Hawaiian animals and plants are isolated from one another not only by virtue of living on different islands. Hawaiian life has also been divided within islands. Recurrent lava flows have constantly reshaped and resculpted the landscape, carving up forests and isolating communities. The result is that each island is made up of many smaller landlocked islands: enclaves of life separated by solid seas of lava.

Both the islands and their landscapes have been characterized by a continual cycle of birth and death. For Hawaii's inhabitants, migration between islands, and between enclaves within islands, has been a necessary way of life. If, as biologists believe, island-hopping is conducive to the origin of species, then it is little wonder that Hawaii, a community of islands in more ways than one, has become one of the world's speciation hot spots.

Of course, there are probably many other factors that have contributed to Hawaii's bumper crop of species. The rich soils of volcanic ash; the mountains that create altitudinal extremes

and produce striking variations of climate; the tropical location—these and many other factors have helped to create the
diverse habitats on display and provide a fertile environment
for speciation. But it is the opportunity for island-hopping
that has probably contributed most to the evolutionary history
of Hawaii's fruit flies.

Island-hopping can produce chance changes in genetic profiles. But this kind of genetic restructuring is not unique to
founding events. Chance genetic changes can occur in any
population that suffers a drastic reduction in size. If disease or
an environmental catastrophe, for example, reduces a population to a fraction of its former size, then the genetic profile of
the survivors may, by chance, be quite different from the one
before the crash.

Humans may have suffered from a "population bottleneck" in
their recent evolutionary history. What else can explain why a
large fraction of our ancestral genetic diversity has gone AWOL?
If you compare human genetic profiles with those of chimps and
gorillas, our closest living relatives, you can turn up some alarming statistics. Try this for starters: there is more genetic diversity
in one social group of chimps in Africa than there is within the
entire human race. In the language of shoe-speak, introduced
earlier, it's as if chimps are walking around with the entire range
of Hush Puppies, while all we have to boast about are some
brown slippers and a grubby pair of flip-flops.

What the profiles suggest is that sometime in the six million
years since we last shared a common ancestor with chimps,
human populations suffered a catastrophic reduction in size.
Nobody has a clue what caused the population to crash. Was it
war? Famine? Plague? Stupidity? Whatever the cause, maybe
the bottleneck was the kick in the backside we needed—a

genetic change of direction—to rouse us from years of monotonous hunting and gathering and make us start thinking about more lofty issues.

Like flies.

Whatever their role in human evolution, population bottlenecks certainly seem to have been a creative force in the evolution of fruit flies. The endemic Hawaiian fruit flies are totally unlike those found elsewhere in the world. They are bigger, bolder, and just a touch more brash than many of their distantly related continental cousins. One group, the "picture-winged" species, could even be described as beautiful, with their large translucent wings, carrying a range of subtle markings and pigmented designs.

All the books (including this one) will tell you that there are about a thousand species of fruit fly on Hawaii. While I do not doubt this figure, I have to confess that in my two weeks on the islands I didn't see a single species in the wild. I did, however, see plenty in the laboratory, including the bizarre *Drosophila heteroneura*. This faintly ludicrous fly has an elongate head, a bit like the head on a hammerhead shark. The weird head is a male-only thing and seems to be a kind of fruit fly equivalent of the peacock's tail, which the male uses to attract females during courtship.

Courtship in Hawaii, like anywhere else, is a big part of a fruit fly's life. But the courtship rituals on Hawaii seem more flamboyant than elsewhere. Males spend a lot of time strutting their stuff on fiercely defended territories. Chat-up routines vary from species to species, but one strategy common to the courtship of many is the pulsating anal droplet. In this strange, yet surprisingly successful, seduction technique, the male arches his abdomen over his back and pulsates an anal droplet of fluid in front of a female's face. If she likes what she smells (and, presumably, is impressed by his anal dexterity), then matters can be taken further. But if a female is underwhelmed, she lets him know by sticking her backside into the male's face, and blasting him with a burst of noxious pheromone.

As I said, everything on Hawaii is done with a touch more style.

As a breeding ground for the origin of species, Hawaii is unsurpassed. Its unique combination of features—isolation, ephemeral landscapes, and environmental contrasts—has proved crucial to its success in the speciation stakes. That, at least, is the theory.

Of course, I should have stated from the start that all this is hypothetical. It's difficult to study speciation, in real time at least. Speciation seems to be a notoriously sluggish affair that far outlasts the lifetime of the average evolutionary biologist. On Hawaii, species seem to have evolved faster than most. The Big Island, for example, has dozens of unique fruit fly species and is only half a million years old. But even this rapid rate is too slow to record in real time.

Because of this speciation go-slow, biologists from Darwin onward have had to develop theories of speciation using indirect evidence from the fossil record, from patterns of geographic distribution, and from subtle comparisons of the genetics, ecology, and behavior of closely related species.

The secondary nature of these observations has always left considerable scope for interpretation and disagreement. Disputes over diverse theories bring together some of the biggest biological egos in the business, each with their pet theory to promote, and each desperate to stake their claim on the ultimate prize in evolutionary biology: to solve the riddle of speciation and inherit the Darwinian mantle.

The nature of the evidence has also given creationists the opportunity to deny that evolution can create new species at all. Of course, should the creationists turn out to be correct, it's great news for the fruit fly. If the evidence on Hawaii is anything to go by, God clearly loves the fly. Why else would he grant the islands one thousand species? On this evidence, he has probably set aside a special place in heaven, a quiet corner of paradise replete with rotting fruit and vegetables, where the flies can live in peace, without fear of spider or lab coat, for all eternity. Amen.

While creationists have stuck to their principles, biologists have seen theories of speciation evolve and change. As the sum total of biological knowledge has grown, the questions surrounding speciation have become ever more sophisticated. Do populations need to be geographically separated from one another to evolve into new species? What is the role of natural selection in speciation? And how many genetic changes are needed for the origin of new species? These—and a hundred other questions—are the currency of today's speciation debate.

Yet, ironically, modern studies of speciation are still hampered by confusion and disagreement over the meaning of the term "species." Clearly, if nobody can agree on what a species is, then there is not much hope of establishing any consensus on how new species evolve.

The species question places Darwin and Dobzhansky, two of the biggest names in evolutionary biology, in direct opposition. At the simplest level, both men agreed that species were different "kinds" of things. Philosophically, however, they were poles apart. They may have written books with almost identical titles, but the word "species" in Darwin's *On the Origin of Species* has a quite different meaning from the one in Dobzhansky's *Genetics and the Origin of Species*.

In Darwin's view, a species was a group of similar individuals, whose boundaries were defined by the subjective hand of a biologist. In *On the Origin of Species*, he wrote:

> In short, we shall have to treat species in the same manner as those naturalists treat genera, who admit that genera are merely artificial combinations made for convenience. This may not be a cheering prospect, but we shall at least be freed from the vain search for the undiscovered and undiscoverable essence of the term species.

To Darwin, species—like the other taxonomic categories of genera, families, orders, and so on—was a useful term in the sense that it helped organize the natural world, albeit along arbitrary lines.

If species were arbitrary, then so was speciation. As far as Darwin was concerned, there was no absolute distinction between the origin of species and the origin of population differences. Terms such as "race," "variety," and "subspecies" could be used to indicate progressive degrees of population divergence. But Darwin considered these words, like the word "species" itself, to be arbitrary and relative terms—abstract boundaries we impose on a natural world always in transition.

In the 1930s, seventy years after Darwin, Dobzhansky came up with an entirely different notion of species. Dismissing Darwin's idea, he argued instead that species were real biological units with their own unique properties. Species, he believed, had intrinsic barriers which prevented them from interbreeding with other species. This idea of species as self-contained reproductive units still dominates biological thinking today.

The notion of barriers to interbreeding was not new. But Dobzhansky was the first to interpret the idea in terms of genetics. He began to see species not as vague and arbitrary collections, but as clearly delineated groups of individuals defined by their ability to mate and freely exchange genes with one another. To Dobzhansky, the movement of genes between populations was the lifeblood of a species, the biological glue that maintained species integrity.

Dobzhansky did not just pluck these ideas out of thin air. He would have got nowhere without biological assistance from the fly—who else? It was fruit flies that helped to convince him that reproductive incompatibility was the line that divides species.

Out in the wild, hybrids between different species of fruit fly were virtually impossible to find, despite intensive searches. When different species of fly were introduced to one another in the laboratory, they rarely got along. Most of them refused to entertain the idea of cross-species sex together. For those with more of an adventurous streak, the results invariably ended with disaster. Hybrid offspring either failed to material-ize, or else they emerged dead, deformed, or sterile. Even pop-ulations of flies that looked almost identical could be isolated from one another by an unwillingness or inability to breed. In short, fruit flies promoted the idea that sexual and genetic incompatibilities, not looks and behavior, were what drew the boundaries between species.

A genetic picture began to emerge of how hybrid sterility could evolve as a by-product of more general evolutionary divergence. Imagine, for example, a single population of fruit flies happily going about its business on a pile of rotting fruit.

One day, a faction of flies decide they have had enough of this particular patch and head off to make a new start somewhere else. Over time, the genetic profiles of the two populations become increasingly different as chance changes, the input of new mutations, and the effects of natural selection all conspire to send them on distinct evolutionary paths.

After a while, the two populations come back together again and indulge in a bit of hanky-panky. Having spent some time apart, hybridization throws together genes that are not used to one another's company. Genes, like footballers, are usually team players. They work in harmony with the genes they have evolved with. So genes that look good and work fine in one team are often useless when transferred to another team.

Dobzhansky saw these genetic incompatibilities as the defining basis of species and speciation. In 1935, he wrote

> . . . species represents that stage of evolutionary divergence, at which the once actually or potentially interbreeding array of forms becomes segregated into two or more separate arrays which are physiologically incapable of interbreeding.

In the 1940s, Dobzhansky's idea was taken up enthusiastically by the German-born biologist Ernst Mayr, who repackaged it under the name of the "biological species concept." The concept had instant appeal because it transformed speciation into a precise and definable event. Speciation became the stage at which two diverging populations are reproductively isolated from one another, when they cease to exchange genes and become genetically independent.

Studies of speciation acquired a focus they had previously been lacking. In this new climate of clarity, biologists turned their attention to "isolating mechanisms"—biological characteristics that formed barriers to the movement of genes between species. Hybrid sterility was, perhaps, the most obvious barrier to gene exchange. But any aspect of a species' biology that prevented the sperm of one species forming a successful partnership with the egg of another came under the broad banner of "isolating mechanisms."

Fruit flies, of course, were at the forefront of this new wave of biological optimism. Differences in the size and shape of the genitals, in the chemical composition of an anal droplet, in the color and pattern of a prominently displayed wing, in the preference for sex at particular times and places—these were just some of the many characteristics that came in for scrutiny.

One of the most celebrated studies of reproductive isolation involves our old laboratory friend *Drosophila melanogaster* and a close relative, *Drosophila simulans*. Despite their similar appearance, *D. melanogaster* and *D. simulans* can be distinguished by a subtle difference in their courtship song. In both species, the male's song is made up of pulsing beats with an oscillating tempo. The only difference between the songs of the two species is the rate at which the tempo changes. It takes a *D. simulans* male about thirty-five seconds to complete a cycle—to go from slow to fast, and back to slow again—whereas *D. melanogaster* males take a relatively sedate fifty-five seconds to complete their song cycle.

Females are extremely fussy about these slight differences in tempo, and have an explicit preference for the song of their own species. Play a *simulans* song and you can guarantee that it will get *simulans* females going; *melanogaster* females, meanwhile, will remain unmoved. But slow the song down a bit, to a more mellow *melanogaster* groove, and the roles are reversed.

The *simulans* females lose interest while the *melanogaster* females get into the mood.

Remarkably, the difference in song tempo between the two species has been traced to a single gene, the *period* gene. A short stretch of DNA is all that distinguishes a fast and slow song cycle. The discovery caused some biologists, in a sudden and uncharacteristic attack of the scientific sound bite, to hail *period* as a "speciation gene." A mutation in the *period* gene was the switch that shifted the tempo, and one species became two as diverging musical tastes became a barrier to gene exchange.

Or so the story goes.

Dobzhansky's ideas about species and speciation have had an immense influence on biology. Go to any high school today and you will find biology teachers drilling the biological species concept into young and impressionable minds, to take up space alongside Newton's laws and a passage from Shakespeare.

But did Dobzhansky get it right? Or was his worldview blinkered by his subjects? As popular and as influential as his ideas have proved, the fact is that had he chosen to work with something other than fruit flies, such as butterflies, marine ducks, or coral reef fish, he might well have dreamed up something entirely different.

Dobzhansky's observation that different fruit fly "species" were sexually incompatible was crucial to the development of his new species idea. But what is true for fruit flies is not necessarily true for nature as a whole. Since Dobzhansky's day, biologists have accumulated a great deal more information on the incidence of hybridization in the wild. And it makes uncomfortable reading for advocates of the biological species concept.

Current estimates suggest that at least 10 percent of animal "species" hybridize in nature. For specific groups, this figure can be much higher. About 40 percent of marine ducks, for example, hybridize in the wild. What is more, many of the hybrids are neither sterile nor deformed, but perfectly capable of interbreeding with either of the parental species.

Of course, you won't hear much about these hybrids, or find references to them in the guidebooks. Ever since Dobzhansky, "hybrid," like "mutant," has become a bit of a dirty word. Fertile hybrids challenge the appealing notion of a species as a cohesive reproductive unit, so they are swept under the carpet and conveniently forgotten.

Dobzhansky's species definition is what Richard Dawkins would call a "meme," a persuasive idea that spreads rapidly through the population. The simple criterion of reproductive isolation atomized the natural world into neat and tidy packets. So popular was this species idea that many biologists believed Dobzhansky had stumbled across a new law of nature. What had been an arbitrary taxonomic category was turned suddenly into a concrete biological reality.

A whole new biological vernacular grew up around this species vision, helping to reinforce the notion of species as real, self-contained entities. Species had "genetic integrity," which was "protected" from "contamination" by "reproductive isolating mechanisms." Is it simply coincidence that this more rigid view of species, with its loaded language, caught on at a time when pure-race ideals and fascism were becoming popular in Europe?

Whatever the historical origins and influences, Dobzhan-sky's species idea was, deliberately or unwittingly, an idealized view of nature. In the end, his definition was no more realistic than Darwin's. Species are reproductively isolated populations, but only if that's what you want them to be.

A philosophical misconception is forgivable. But the biological species concept is not even a very useful way of describing the diversity of the natural world. Dobzhansky believed that reproductive isolation was an objective criterion of distinction, and far better than Darwin's woolly view of species as different "kinds" of things. Of course, in many cases the two definitions amount to the same thing: different "kinds" of things are also reproductively isolated from one another. But this is not always the case. Different "kinds" of things can remain distinct, despite freely interbreeding with one another. To find a good illustration of this, look no further than the Galápagos Islands, to those icons of evolutionary biology, Darwin's finches.

Most people would think of these fourteen species of rather drab birds as good biological species, in the Dobzhansky sense. But they are not. They freely interbreed to produce fertile hybrids. So why, you may ask, do the species differences not erode? Why do we continue to see distinct groups of individuals? The reason is that hybrids have beak shapes intermediate in size between the two parental species. Although the hybrid birds are physically fit and healthy, their beaks make inappropriate feeding tools for the seed types on the islands.

The situation, however, is not static. During the 1980s, a period of exceptionally heavy rain triggered temporary changes in the islands' vegetation. There was a shift in the distribution of seed types. Suddenly, the beaks of the hybrid birds were better adapted than those of the parental species, and species differences began to erode. Distinctions may reappear if and when the vegetation returns to its former state.

Darwin's finches show how an obsession with reproductive isolation can make us blind to the subtleties of evolutionary change. Reproductive isolation is only one state of evolutionary divergence; there is no logical reason why it should assume absolute importance. We consider ourselves to be good biological species because we cannot interbreed with chimps and gorillas, our closest living relatives. But we have no idea when this reproductive isolation occurred. Perhaps it took place long after we became physically and behaviorally distinct. If it was biologically possible to interbreed with chimps today, would we still consider ourselves separate species?

The great irony is that the world is actually a much more simple and satisfying place without Dobzhansky's species definition. If evolution teaches us anything, it is that life on Earth has always been in transition. So why try to pin it down with rigid formulas?

When we take a snapshot of this changing world, we do not see a complete continuum of forms. There are noticeable gaps that bring clusters of similar individuals into sharp relief. In an abstract sense, we can think of these clusters as peaks on an undulating three-dimensional surface. This surface, like the landscape of Hawaii, comes complete with mountains, hills, and valleys. Peaks on the surface represent clusters of individuals sharing similar genes. The valleys equate to rarer combinations of genes, the kinds normally found in hybrids. In this scheme, clusters are clearly self-evident. But deciding whether these clusters are varieties, races, or species is an arbitrary exercise akin to deciding whether a peak is a hill or a hummock.

Like the Hawaiian landscape, the genetic landscape is not

fixed and static. Over time, evolution keeps the surface in a state of constant flux. Population bottlenecks cause small hills to split off from the base of larger peaks. Natural selection can make mountains out of molehills, and hybridization can reduce twin peaks to plateaus. Ever since life began, this landscape has been endlessly reworked and remolded. Little wonder, then, that it has been so reluctant to succumb to all-embracing definitions. Darwin's definition of species was indeed woolly, but only because nature is made that way.

Of course, you can't really blame Dobzhansky for imposing a dubious concept of species on the world. If you're looking for the real ringleader, then blame the fly. It is the fly that helped to give us a perennially popular, if slightly skewed, vision of nature. But considering the many good things the fly has done for us, we shouldn't begrudge it the odd mistake.

8

TIME'S FLY

On Manhattan's Fifth Avenue, between Thirty-second and Thirty-third Streets, and a few doors down from the Empire State Building, is a small men's hat store called "JJ Hats." It's a charming place, all creaking drawers and wood-framed cabinets. The shop doesn't look as if it has changed much since it first opened its doors, almost a hundred years ago. Step inside and you could be in New York, circa 1910.

This shop dates back to the era of Thomas Hunt Morgan and his early experiments with the fly. Who knows, Morgan may even have visited the store, making the three-mile trip down from the Columbia University campus at Morningside Heights. Today, I'm making that journey in reverse. I'm getting kitted out with some suitably historic headgear for my pilgrimage north, to Morgan's Fly Room, the place where the fly's scientific success story began.

But first I have to decide what kind of hat to buy. There's no shortage of choice at JJ Hats. Every conceivable style and color is available. I'm tempted by a Stetson or a fedora, but eventually plump for something a bit more modest, a "newsboy," a

hybrid between a big floppy beret and a peaked cap. The newsboy was popular in the 1920s and 1930s, and resurfaced in the
1970s as the headgear of choice for drug dealers and pimps.
The hat looks faintly ridiculous on me, but outside on the
street nobody seems to notice. This is New York, after all; if you
can't look ridiculous here, then where can you?

With the newsboy fixed at an implausible angle, I head north
up Fifth Avenue and into Central Park. It's a beautiful, sunny
morning in February. The park is alive with rollerbladers, uni-
cyclists, jugglers, and mime artists, all out and about on their
Saturday morning pose. The newsboy blends in perfectly.

Leaving the skyscrapers of Midtown Manhattan behind, I
wander through the park, over toward Central Park West. The
Lake is still frozen over, and its mirrored surface reflects a bril-
liantly focused light on the lavish architecture of the Upper
West Side. The Dakota, in particular, looks stunning. Gripped
by its gothic opulence, I'm suddenly lost to a moment of
reverie, to melancholy daydreams of John and Yoko and a
youth that all seems so long ago.

Thoughts of the Fly Room bring me back down to earth. I
reset the newsboy and continue north, up the west side of the
park, past the American Museum of Natural History. Pricey
apartment blocks eventually give way to a more down-to-
earth, urban ecology. At the northwest corner of the park, I
turn left down Cathedral Parkway, and then right up Amsterdam Avenue, climbing the gentle slope that leads me to the
rear gates of Columbia University.

Unfortunately, I've forgotten the room number of the Fly
Room. But I'm not too worried. I know that the room was in
Schermerhorn Hall, which I can easily find on a map. What's

more, I'm convinced that there will be plenty of signs guiding visitors toward this world-famous laboratory. I don't know whether the Fly Room still exists in its original form. But even if it doesn't, I feel sure that there'll be some kind of plaque nearby to commemorate its place in scientific history; maybe even a simple museum.

I find Schermerhorn Hall with little difficulty. It is a pleasant, if unremarkable, multistory building with some neoclassical touches here and there to liven up a plain nineteenth-century facade. I look around for the signposts pointing toward "Morgan's Fly Room." There are none to be found, not on the outside, at least.

Inside, the place is deserted. Okay, it is a Saturday, but I wasn't expecting the place to be quite so empty. Don't academics work on the weekend? I take the lift to no floor in particular and roam the whitewashed corridors, carefully inspecting each door for any hint or clue, all the time struggling to remember the room number.

I am beginning to lose patience when I bump into a couple of anthropologists.

"Sorry to bother you. I wonder if you can help me? I'm trying to find Morgan's Fly Room."

Blank looks.

"You know, the fruit fly, *Drosophila melanogaster?*"

More blank looks. Perhaps the newsboy is putting them off. I quickly remove it.

"It was in this building—here at Columbia University—Thomas Hunt Morgan—he made the fruit fly famous."

No improvement. It wasn't the newsboy.

After a little more explaining, I discover that, yes, they have heard of the fruit fly, but no, they don't know anything about

Morgan or the Fly Room. So I continue my search for a little while longer, scouting the stairs and corridors, hoping to catch the scent of genetic history.

Slowly, it begins to dawn on me that the Fly Room is not quite as legendary as I'd thought, and after half an hour of fruitless searching, I leave the building, disappointed. On the way out of the campus, I make some further inquiries. Nobody I ask can help me. Clearly, the Fly Room is not one of Columbia University's top tourist attractions.

When I returned to the U.K., I e-mailed Jim Erickson, a fruit fly biologist at Columbia University. I wanted to make sure I hadn't overlooked anything, and so I asked him about the current status of the Fly Room. He confirmed my suspicions. The Fly Room no longer exists in its original state. In fact, a new wall has been built since Morgan's era, reclaiming part of the old room as a hallway. There is no museum and no commemorative plaque.

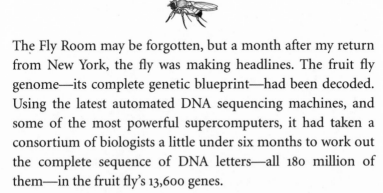

The Fly Room may be forgotten, but a month after my return from New York, the fly was making headlines. The fruit fly genome—its complete genetic blueprint—had been decoded. Using the latest automated DNA sequencing machines, and some of the most powerful supercomputers, it had taken a consortium of biologists a little under six months to work out the complete sequence of DNA letters—all 180 million of them—in the fruit fly's 13,600 genes.

The journal *Science* published a bumper issue to mark the occasion. Its cover featured a male and female fruit fly, superimposed on rows and rows of A's, G's, T's, and C's—the letters of the

DNA alphabet. Inside, there were foldout charts, earnest reviews, and considered opinions from a host of big-name scientists.

So where does the sequencing of the fly genome feature in the list of the all-time great fruit fly discoveries? There's no doubt that it was an extraordinary achievement. But it is more a triumph of technology than of science. In many ways, it can be seen as the culmination of a long mapmaking tradition that goes all the way back to Morgan's Fly Room. The technology may have changed beyond all recognition, but you can trace a direct line of ancestry from today's automated DNA sequencers to the milk bottles and wall charts of the Fly Room.

When it comes to genetic mapmaking, Morgan, Sturtevant, and Bridges were the original pioneers. Using controlled crosses between mutant flies, they came up with an ingenious way of working out the linear order of genes along the fruit fly's chromosomes. It was simple but revolutionary technology. Everyone wanted a taste of it, and the world went mapping crazy.

Fast-forward to the 1970s, to the arrival of DNA sequencing and a new era of mapmaking. Instead of the linear order of genes, biologists now had the tools to work out the linear order of letters within a sequence of DNA. Once again, the arrival of a powerful new technology had biologists in a frenzy.

Ironically, both these technologies—genetic mapmaking and DNA sequencing—rekindled the flames of the naturalist tradition. It might seem implausible that twentieth-century geneticists could have much in common with the naturalists of the nineteenth century. But genetics, a science that evolved out of a rejection of the naturalist philosophy, has returned again and again to embrace it.

The naturalist philosophy was built on the tenets of observation and description. In simple terms, naturalists are biological stamp collectors. In the nineteenth century, these stamps came in the form of millions of preserved animal and plant specimens, which were collected, described, and classified. By the twentieth century, the stamps had changed—they were now genes or DNA sequences—but the basic approach to biology remained the same. In the 1920s and 1930s, pinpointing a gene on a chromosome map was the high-water mark of many a biological career. And in the 1970s and 1980s, the dream of deciphering a DNA sequence was the only reason that many molecular biologists got out of bed in the morning.

The new generation of genetic naturalists embraced novel technologies and experimental techniques, but only to enhance or refine their observations and descriptions, in the same way that a naturalist would use binoculars to identify a bird, or a microscope to probe the details of an amoeba. Hypothesis testing through carefully controlled experiments—the foundation of the experimental philosophy—went out the window, as legions of biologists contented themselves with either mapping or sequencing genes.

Natural history in the nineteenth century was all about the three C's: collecting, cataloging, and classifying. What lay behind the exhaustive observations and compulsive note-taking was the desire to uncover universal patterns and relationships among living things. Whether these patterns emanated from the divine hands of the Creator or the blind forces of evolution, the formalities of classification remained the same. Leaves still had to be drawn, legs still had to be counted, and beaks still had to be measured. Natural history museums, such as those in New York and London, were built to house vast cities of preserved specimens, and stand as monuments to the Victorian obsession with order.

These old museums have a modern equivalent, which you can find by logging onto the Internet. The Victorian naturalist tradition lives on in the computer databases that catalog and classify the genetic maps and the thousands of DNA sequences that have been collected over the years. Inside these electronic museums, you can find all manner of genetic artifacts. Here, too, you will find the complete DNA sequence of the fly, the yeast, the nematode worm, and a host of bacteria. You can also find one of the most recent additions: the three billion DNA letters that make up the complete human blueprint.

You're free to browse around the museum. Take a look inside the fruit fly gallery, for instance, and you might come across something that looks like this:

```
AATTCGCCGAATATGCCGTACGTCGATTAACGCTCT
TAGCTTACTACGTCATACTGGGTATACTCACGGAG
TAATCCGTACGTACGTACGTCATCGTATACGTACGT
TATCGCTACTGCTCGT ...
```

Riveting, isn't it?

Have a peek inside the yeast gallery and you might find something like this:

```
GGGCGTAAAATGTTGTGCGCTCTTTACACAGCGTAC
GATCCAAGTACGATTACGTTCATGACTGCGATCAGTAC
CATGGTACGCTACTGCATGCATGGACTACGTACTGGCAT
GCTGCATGGCTGACT ...
```

Feeling enlightened?

On its own, a DNA sequence tells us little except how interminable it is. Just as there's not much insight to be gained from simply counting the number of body hairs on a single insect specimen, there's not much you can say about the hundreds of letters in a DNA sequence. Bring some additional information to bear on top of it, however, and things start to get more interesting. Knowledge of the genetic code means that a DNA sequence can be used to predict the shape and structure of its protein product. In turn, the protein's shape and structure can reveal clues to its role in the day-to-day business of life.

A single DNA sequence is one thing. But bring all the DNA sequences together and the value of the DNA museum becomes more obvious. Just as the Victorians used vast collections of specimens to infer evolutionary patterns between living things, the catalog of DNA sequences can be used to infer evolutionary and functional relationships between genes within and between species.

In one sense, sequencing the fruit fly genome brings this particular bit of natural history to a close. For the fruit fly, the genetic stamp collecting is over. All that remains now is to work out how this enormous inventory of letters puts itself into practice. That should keep biologists happy for the next century or so. In the meantime, we can look back on a remarkable hundred years, on a fruit fly century that has transformed the face of biology.

It all began under William Castle's watchful eye at Harvard University. Those were more innocent times. Biologists were still arguing about all manner of conundrums: the credibility of Darwinian evolution, the physical nature of heredity, and the most effective way of getting Hugo de Vries's mammoth book *Die Mutationstheorie* back home without suffering personal injury.

Biology has come a long way since then, thanks in no small

measure to the fly. Its curriculum vitae reads like a checklist of the twentieth century's biological landmarks: the foundations of genetics, the fusion of genetics and evolutionary biology, the genetic dissection of behavior, embryonic development, and aging—these are just some of the fruits of a century's hard laboratory labor.

The impact of these discoveries goes way beyond the narrow confines of fruit fly biology, and therein lies the secret of the fly's success. The fly has proved itself to be a biological beacon, throwing light on universal rules of life. Again and again, new discoveries in fruit flies have led to parallel discoveries in many other forms of life, humans included.

Take embryonic development, for instance. When genes that control the building of a fruit fly body were discovered in the 1970s, it was a remarkable breakthrough. For the first time, biologists had a glimpse of how the journey from egg to embryo was orchestrated and controlled—in flies, at least. But soon, similar genes were turning up in sea slugs, frogs, mice, and humans. Far from being unique, the fruit fly development plan was a guide to development in other forms of life.

We now know that flies and mammals do not just share genes that orchestrate the building of the basic body plan. They also share genetic switches that turn on the development of eyes, limbs, nerves, and hearts. In fact, so similar are some of these genes that they are interchangeable. You can knock out the gene that controls eye development in a fly and replace it with the corresponding gene from a mouse, and the fly will develop eyes as normal.

Since its success in developmental biology, the fly has pinpointed many more genes and biochemical pathways that are conserved across the animal kingdom. From birth to death, the fruit fly has uncovered a host of well-trodden genetic trails, highlighting the remarkable economy of evolution.

It is this history of success that gives biologists the belief that the cause of human alcoholism or drug addiction can be found in the behavior and genetics of an intoxicated fly; that solutions to jet lag and sleeping disorders can be discovered in heads rolling off a fruit fly guillotine; that cures for memory loss and trauma can be unearthed in the simple training of a fruit fly learning exercise; and that the secret of eternal youth lies in the aging antics of a fruit fly Methuselah.

Of course, the fly has not had everything its own way. Throughout the twentieth century, it has had to endure repeated challenges from a host of laboratory competitors. Today, the fly faces stiff competition from new challengers like the mouse and the nematode worm, *Caenorhabditis elegans*. Both animals have their admirers. The mouse, being a mammal, is often touted as a more suitable model for human biology, but it is less amenable to genetic tinkering, and lacks the durability of the fly. The nematode worm, on the other hand, with its thousand-cell simplicity, has proved its worth in areas such as aging and embryonic development. But the worm's lack of charisma and slim repertoire of behaviors mean that it is hardly a match for a *bon viveur* like the fly.

The fact is that the fly remains the original and the best all-around model organism. Its early laboratory experiences may have been inauspicious, but as soon as it found its way to New York and to Thomas Hunt Morgan's laboratory at Columbia University, it never looked back. Since then, thousands of biologists have been attracted by the fly's enormous talent for putting fresh wind in sagging scientific sails.

The fly may have cemented itself in history. But others have not been so lucky. Spare a thought for William Castle, the man who first took the fruit fly under his wing. In his day, Castle was a high-profile figure in biology. Perhaps his most striking contribution to science came in 1909, with his study on the gonads of guinea pigs, work that dealt a hammer blow to the dwindling band of biologists still devoted to Lamarckian evolution. And yet Castle, the original fruit fly pioneer, is now almost totally forgotten. Nobody, besides a few historians of science, remembers him.

Perhaps the source of this collective amnesia can be traced to a remark Castle made in 1919, in one of his own scientific papers:

> That the arrangement of genes within a linkage system is strictly linear seems for a variety of reasons doubtful. It is doubtful, for example, whether an elaborate organic molecule ever has a simple string-like form.

Now, it is easy to be wise after the event, but in the space of these thirty-six words, Castle makes two of the most erroneous scientific predictions of all time. Not only had Morgan more or less shown that genes were linearly arranged on chromosomes, but, in 1953, Watson and Crick discovered that the genetic material—DNA—had a simple stringlike form. To put things into perspective, Castle's forecast was about as accurate as the one made by the A&R man at Decca Records who, in 1962, rejected the Beatles, proclaiming: "Groups of guitars are on the way out...."

But scientific faux pas alone cannot explain Castle's disappearance from all but the most obscure texts of biological history. After all, Morgan is the man who at one time rejected Mendelian genetics, the chromosome theory of inheritance, and Darwin's theory of evolution by natural selection. Not a bad collection for a future Nobel Prize winner.

No, the fact is that few scientists leave a long-lasting imprint on the memory, as my visit to the Fly Room ably demonstrated. Even if you believe that your work is pushing back the frontiers of science (and, let's face it, there are many out there who do); even if you are on first-name terms with the editors of all the top scientific journals; and even if you make it onto the board of some prestigious grant committee, the chances are that posterity will reduce your lifetime's work to a single-sentence footnote in the annals of some obscure scientific biography. Just ask William What'shisname.

The fruit fly, meanwhile, lives on.

FRUIT FLY FACTS AND FEATS
A Collection of Fruit Fly Odds and Ends

Below is a brief list of mutants that did not find their way into the main part of the book but deserve a mention nonetheless.

chico With fewer and smaller body cells than normal, this diminutive fly (the name means "small boy" in Spanish) is less than half the normal fruit fly size. Popular with Japanese bonsai lovers, apparently.

pirouette Obsessed with the geometry of circles, this fly starts its adult life tracing big wide arcs around its cage. Gradually, the fly's turning circle becomes ever smaller until it enters a phase of balletic pirouetting. Either through starvation, advanced motion sickness, or a combination of the two, the fly finally collapses and dies, in tangential peace.

shaker All shook up it may be, but with its convulsive twitches and arrhythmic leg-shaking, this is no fruit fly Elvis. A dead cert for a short adult life.

dachshund It's definitely not a dog's life for dachshund. The fly's stubby little legs are totally useless for walking, but this doesn't seem to stop it trying. It flails about feebly for a few days before succumbing to dehydration and death.

drop dead A tragedy waiting to happen, drop dead begins life in rude health, enjoying a normal upbringing as a maggot and pupa. Even in the early days of adulthood there are no outward signs of the trouble ahead. Everything seems so normal, almost disturbingly so. And then, suddenly, and without warning, the fly totters on its feet and drops dead. *C'est la vie.*

eagle A fly that seems to have ideas way above its station, the *eagle* mutant holds its wings out at right angles to the body, as if it is dreaming of a bigger and better life, soaring majestically, high on updrafts of air, the master of all it surveys. Or perhaps it just has a dodgy set of wings.

forkhead With various bits of its head growing in places where the gut should be, it's a safe bet that this homeotic mutant suffers from acute indigestion.

groucho Bushy eyebrows are what links this fly with the chattering Marx brother. A pity, really, because a fly with a talent for witty one-liners would really be something.

van gogh An evocative name for what is, in all honesty, quite a boring mutant. The hairs on the fly's wings form swirling patterns rather than the more regular arrangement found in normal flies. The biologists who named the fly decided that these patterns "bring to mind the swirling brush strokes the artist used in some of his paintings." That's impressionism for you.

genghis khan A mighty fruit fly warlord that rapes and pillages its way across continents? Not quite. The name comes, in part, from the accumulation of actin—a protein that gives power to muscles—in the mutant egg. Still, *genghis khan* could yet set a trend for more fruit fly dictators. Anyone for *adolf hitler, benito mussolini,* or *francisco franco*?

TEN THINGS YOU NEVER KNEW ABOUT FRUIT FLIES— AND PROBABLY NEVER WANTED TO

1. The literal translation of *Drosophila melanogaster* is "black-bellied dew-lover." Although the "black belly" part makes sense—it refers to the black tip of the male fly's abdomen— "dew-lover" seems a bit wide of the mark. Perhaps this slight oversight is evidence that Carl Fredrik Falleén, the Swedish entomologist who first named and described *Drosophila* in 1823, had spent too much of his time around breweries and vineyards with his subjects.

2. At various times, the fruit fly has been known colloquially as the vinegar fly, the pomace fly, the wine fly, the banana fly, and the pickled fruit fly.

3. Some fruit fly species do not care for fruit at all but prefer to lay their eggs in mushrooms, cacti, or flowers. There are even a few bizarre species that have abandoned plants altogether for a more harum-scarum lifestyle. *Drosophila carcinophila,* for example, lays its eggs in a land crab's "nephritic groove"—the crab's own onboard urinal. Once hatched, the young fly maggots feast on the crab's excrement.

4. "Reactions of the pomace fly" was the rather nebulous title to a 1905 paper by F. W. Carpenter, published in the journal

American Naturalist. An otherwise unremarkable piece of science, it holds the distinction of being the first-ever published laboratory study of the fly.

5. Most *Drosophila* species can mate in the dark, using their species-specific song to find the right mate. But *Drosophila subobscura* is a rare exception. Being one of the few species that lack a song, it relies on visual cues during courtship and insists on having the lights on during sex.

6. *Drosophila pseudoobscura* has a transparent scrotum. Of course, in the strictest sense insects don't have scrotums—they're a mammalian thing. But the fly has something similar—a sheath that covers the testes. In *D. pseudoobscura* this sheath is transparent and provides a clear view of the bright orange testes below. For this reason, *D. pseudoobscura* has always been valued in studies of hybrid sterility. Testes size provides a good measure of fertility. The bigger the balls, the more potent the sperm. Transparency means there is no need for messy dissections. You can measure the fertility of a fly just by whipping it onto its back and taking a quick glance at the gonads.

7. *Drosophila bifurca* produces sperm measuring over 58 millimeters in length—more than ten times the length of its body.

8. While the fruit fly head has its uses, there are occasions where it seems surplus to requirements. There is evidence, for example, to show that flies are better at learning some kinds of tasks without a head. To demonstrate, take a fly, fix it to a stick, and suspend it above a bowl of salt water. Next, tie a short stretch of fine wire around one of the fly's legs, so that the wire can just touch the surface of the liquid. The fly will not be very

happy about this state of affairs and will wave its legs around in protest. But each time the wire brushes the surface of the liquid the fly receives a small electric shock. Remarkably, decapitated flies learn to keep their leg up—and avoid the electric shock—better than flies with their heads intact. This experiment, first performed in the 1970s, not only shows that heads can be a distraction, it also illustrates how learning can take place in nerves outside the brain.

9. Fruit flies can become addicted to crack cocaine. Under the influence of the drug, the flies indulge in manic grooming. At high doses, they walk backward, sideways, and in circles. Repeated use induces the reverse tolerance to the drug that is typical of human addicts.

10. One pair of flies can easily produce two hundred offspring in a fortnight. If every one of those flies and all their descendants continue at that rate of productivity, then by the end of a year, you'd have about one billion billion billion billion billion billion billion flies.

SELECTED READINGS

Robert Kohler's *Lords of the Fly: Drosophila Genetics and the Experimental Life* (University of Chicago Press, Chicago, 1994) and Garland Allen's *Thomas Hunt Morgan: The Man and His Science* (Princeton University Press, Princeton, 1978) were invaluable companions throughout the writing of this book. Both are excellent in recounting the atmosphere and experiences of the early fly years. For a more general history of American biology, try *The American Development of Biology* (Ronald Rainger, Keith R. Benson, and Jane Maienschein, eds., Rutgers University Press, London, 1991), or Jane Maienschein's *Transforming Traditions in American Biology, 1880–1915* (Johns Hopkins University Press, Baltimore, 1991). Franklin Portugal and Jack Cohen's *A Century of DNA* (MIT Press, Cambridge, 1977) provides a good general overview of the history of genetics.

For more information on Hermann Muller, see Elof Axel Carlson's *Genes, Radiation, and Society: The Life and Work of H. J. Muller* (Cornell University Press, Ithaca, 1981). Though tending toward the technical, Peter Lawrence's *The Making of a Fly* (Blackwell Scientific Publications, Oxford, 1992) contains highly readable short histories of some of developmental biology's key moments. "The molecular architects of body design" (*Scientific American,* February 1994, pp. 36–42) by William McGinnis and Michael Kuziora is perhaps a better place to start for a brief introduction to developmental genetics.

Jonathan Weiner's *Time, Love, Memory* (Faber & Faber, London, 1999) is both a biography of Seymour Benzer and an introduction to the genetics of fruit fly behavior. Robert Dudley's "Evolutionary origins of human alcoholism in primate frugivory" (*Quarterly Review of Biology,* vol. 75, no. 1, March 2000) explores the roots of our fondness for alcohol.

The Evolution of Theodosius Dobzhansky (Mark B. Adams, ed., Princeton University Press, Princeton, 1994) is a very readable primer to Dobzhansky's life and work. *The Evolutionary Synthesis: Perspectives on the Unification of Biology* (Ernst Mayr and William B. Provine, eds., Harvard University Press, Cambridge, 1998) traces the origins and influences of evolutionary genetics. Jeffrey Powell's *Progress and Prospects in Evolutionary Biology: The Drosophila Model* (Oxford University Press, Oxford, 1997) is pretty much what it says: a review of the fly's contribution to evolutionary biology.

For more information on the evolutionary intricacies of animal sex, see *The Evolution of Mating Systems in Insects and Arachnids* (Jae C. Choe and Bernard J. Crespi, eds., Cambridge University Press, Cambridge, 1997), John Krebs and Nick Davies's *An Introduction to Behavioural Ecology* (Blackwell Scientific Publications, Oxford, 1993), or *Sperm Competition and Sexual Selection* (T. R. Birkhead and A. P. Møller, eds., AP Professional, London, 1998).

Steven Austad's *Why We Age: What Science Is Discovering about the Body's Journey Through Life* (John Wiley, Chichester,

1997) provides a good introduction to the biology of aging, while Tom Kirkwood's *Time of Our Lives* (Weidenfeld & Nicolson, London, 1999) is a popular and highly readable account of the disposable-soma theory of aging.

For a recent overview of speciation, try *Endless Forms: Species and Speciation* (Daniel J. Howard and Stewart H. Berlocher, eds., Oxford University Press, Oxford, 1998). Those interested in the evolutionary significance of hybrids should see Michael Arnold's *Natural Hybridization and Evolution* (Oxford University Press, New York, 1997). Jonathan Weiner's *The Beak of the Finch* (Alfred A. Knopf, New York, 1994) brings the story of Darwin's finches up to date, while *Species: New Interdisciplinary Essays* (Robert A. Wilson, ed., MIT Press, Cambridge, 1999) is an excellent trawl through the minefield of conflicting ideas about species.

And, finally, for the biologist's take on the sequencing of the fruit fly genome, see the special issue of *Science* (vol. 287, 24 March 2000).

INDEX